GROWTH AND ADJUSTMENT IN NATIONAL AGRICULTURES

# GROWTH AND ADJUSTMENT IN NATIONAL AGRICULTURES

## Four Case Studies and an Overview

Edited by
J. P. O'HAGAN
*Chief*
*Global Perspective Studies Unit*
*FAO, Rome*

Published by arrangement with the
FOOD AND AGRICULTURE ORGANIZATION
OF THE UNITED NATIONS

*First published 1978 by*
THE MACMILLAN PRESS LTD
*London and Basingstoke*
*Associated companies in Delhi*
*Dublin Hong Kong Johannesburg Lagos*
*Melbourne New York Singapore Tokyo*

British Library Cataloguing in Publication Data

Growth and adjustment in national agricultures
1. Agriculture – Economic aspects – Case studies
2. Economic development – Case studies
I. O'Hagan, J P II. Food and Agriculture Organization
338.1'07'22 HD1411

ISBN 978-1-349-03478-9 ISBN 978-1-349-03476-5 (eBook)
DOI 10.1007/978-1-349-03476-5

# Contents

# Preface

The four country studies in this volume have been selected from more than a dozen undertaken in recent years by the Food and Agriculture Organization in support of its preparation of a strategy of international agricultural adjustment. (The full list is given in the Appendix.) Together they cover a wide range of approaches to an issue which all countries face: the need for continuing and sometimes radical structural adjustments in agriculture to adapt to new requirements and potentialities both within the sector and between agriculture and other sectors.

The studies on Poland and the Sudan were prepared for the November 1975 Session of the biennial FAO Conference and those for Japan and Sweden for the preceding session. Apart from editorial changes and minor shortening, the texts remain as presented to the FAO Conferences. In each case, the responsibility for the contents was that of the Director-General of FAO, and the study was prepared under the direction of an officer of this Organization. However, very close contact was maintained with authorities in each country, and the studies reflect and benefit from the intensive discussions which took place on draft texts between the authors, FAO staff, Ministries and others concerned. The introductory chapter, 'Adjustment Processes and Issues in Agriculture', has been written for the present publication.

Three of the four studies presented in this volume deal with developed countries. The main theme of the work which they supported was the feasibility of expanding the flow of agricultural exports from developing to developed countries, including consideration of adjustments in developed countries which would consequently be needed. These research studies were therefore designed first to contribute to a better understanding of the agricultural adjustment process in the varying conditions met in developed countries. The exception is the brief paper on the Sudan which looks at the obverse of this issue: the scope for and implications of increasing agricultural exports in the case of one of the least developed countries.

The four papers have been brought together in one volume because it is believed that they do present information and an interpretation not otherwise readily available on the evolution of structural adjustment in national agricultures. Furthermore, the notion of developed countries absorbing more imports of agricultural products from developing countries which was largely in abeyance during the food crisis years around 1974 and 1975, is now re-emerging as a major issue and will again focus attention on adjustment in developed countries. The contents of this book are therefore of current as well as of historical interest.

# Acknowledgements

The present volume contains studies which were undertaken by the Economic and Social Policy Department, FAO. The contributions of the following consultants are gratefully acknowledged:

| | |
|---|---|
| For the study on Japan | Mr Y. Inouye, Asian Productivity Organization, Tokyo |
| For the study on Poland | Dr A. Wos, Professor and Director, Institute of Agricultural Economics, Warsaw |
| | Dr Z. Grochowski, Professor and Deputy Director, Institute of Agricultural Economics, Warsaw |
| For the study on the Sudan | Dr El Sayed Gaballah, Cairo |

# Adjustment Processes and Issues in Agriculture: An Overview

Agriculture is changing rapidly. Traditional attitudes and techniques are confronted by new social requirements and technological possibilities. A worker on the land in developed countries is likely to use more capital and to have a higher rate of growth of productivity than his counterpart in industry, and in developing countries the farmer is becoming aware of innovations withheld from him only by lack of finance. The area of land under cultivation has been expanded more rapidly in the last decade than was thought possible, and average crop yields have risen more in the post-war period than over virtually the whole of recorded agricultural history; they are, however, usually far below biological limits. The extent and swiftness of recent change in the agriculture of developed countries is without precedent, as is also the potential and the need for agricultural change in developing countries. However, despite the dynamism of modern production technology in farming and the world's need for more food, agriculture is a declining industry in the sense that its share of total output is falling in all countries, reflecting the fact that the income elasticity of demand for farm products is lower than for most other goods and services.

Against such a background there is continuous adjustment, both within agriculture and between this and other sectors, nationally and internationally. The structure of agriculture in terms of numbers of farms, their average area, the employment they can offer and the way in which they are organized has changed substantially in many countries. The percentage contribution of on-farm activities to national income drifts downwards in virtually all countries, although the share of agriculture-based employment is much more resistant. The level and composition of commodity output is continuously and substantially altered as supply and demand interact.

Changes of these kinds impose large and frequently difficult adjustments on farming communities as well as on the country as a whole, their nature varying with the conditions of the country. The case studies in the present volume vividly illustrate the point: in Poland, peasant farming had to finance the industrialization of the country while at the same time adapting itself to new forms of political and economic organizations. As Japanese farming families turned increasingly to non-farm activities to improve their incomes but did not give up their land, the farms had to be increasingly mechanized so that they could be worked by part-time operators and families, a development which also favoured products such as rice which do not need year-round attention. To help to meet the labour requirements of industry, to prevent agricultural output from rising too rapidly, and to integrate the farm sector more fully into the economy as a whole, Swedish policies fostered the simultaneous development of a sustained outflow of labour from this sector and the extensive modernization and enlargement of the remaining farms. To achieve an acceptable rate of economic development, and to sustain that development, the Sudan must expand food exports despite the existence of unsatisfactory nutritional conditions amongst its own people.

The processes and issues of agricultural adjustment can be viewed at three different stages: within farming itself, where they concern especially the size of farms and the internal organization of the sector; nationally, where they arise particularly in regard to the balancing of commodity supply and demand and to the linkages between the farm sector and the economy as a whole; and internationally, where through trade the effects of the adjustments carried out in the prior two stages are transmitted to other countries, where they affect their current economic conditions and prospects for development, including their food supplies. Although the nature of these problems at the different stages is distinctly different, the three stages of adjustment are also closely linked.

### *Adjustment within farming*

The basic task here is to create or maintain farms which are viable economically and socially. A huge number of the world's farms do not measure up to whatever standard of viability would be appropriate to their country. They are too small in terms of production resources per worker or dependant to provide living standards reasonably equivalent to those in other occupations. They are also too small to take advantage of some of the new production techniques and to have adequate access to such services as institutional farm credit.

The basic way in which this issue has been faced in developed countries is for members of farming families to transfer to non-farm employment. Lower incomes in farming and other disadvantages relative to other occupations have provided the push. The growth of manufacturing and service industries in the post-war period has provided the pull for a labour outflow from farming which has been sustained at surprisingly high average rates. The experiences cited in the case studies are typical. In Sweden between 1960 and 1970 there was a net reduction of 43 per cent in persons employed in agriculture. Approximately half transferred to other employment, while the other withdrawals were through retirement on pension, death and disabilities. Agriculture has been and still is the major reservoir of manpower for expanding Polish industry, and it is estimated that over the period 1946–75 the outflow from rural districts totalled over 5 million persons. This migration is expected to continue, and in the decade 1981–90 employment in agriculture in Poland is projected to decline at an average rate of almost 6 per cent a year. Between 1960 and 1971 the share of the Japanese labour force in agriculture almost halved, dropping from 27 to 14 per cent.

Movements of people of these magnitudes require imaginative and well-organized arrangements if the internal migration is to take place without distress, and most developed countries now have retraining facilities, employment advice, monetary grants and so on for facilitating and humanizing the transfers. Such steep reductions in farm labour force directly raise the amount of productive resources available per worker remaining in agriculture, and governments have taken the opportunity to introduce policies of consolidating land into larger units and assisting the restructuring and re-equipping of viable farms. The Swedish case study gives a detailed account of actions undertaken in that country. As shown in the study of Poland, where individual farms typically still remain small, cooperative arrangements can provide pools of productive equipment, such as specialized machinery, above and beyond what the individual farmer could hope to own.

Experience shows that in developed, industrialized countries, structural adjustments that are truly far-reaching in their scope can be very successfully achieved. The extent of flexibility is greater than has traditionally been thought. It may indeed be claimed that if by economic criteria the restructuring of agriculture has seldom been radical and rapid enough to enable agriculture to be so integrated into the national economy that it need no longer be treated as a permanent 'special case' (apart from environmental considerations), from a social viewpoint the rate of change has often seemed excessive. The socio-economic balance of whole areas has been upset, strategic or ecologically important regions have become depopulated, the enlargement of urban populations, already too numerous, has been accelerated, and the average age of those left in farming has risen unduly. These qualifications should not be allowed, however, to detract from the very considerable success there has been in the process of structural adjustment of agriculture in developed countries.

The major impediments are twofold. First is the procedure followed in most countries of raising the income of farmers through price supports.[1] Whatever the political and administrative merits of this approach, it inevitably encourages the retention of excessive or unsuitable resources in agriculture. During the several years when world food shortages loomed, this issue naturally became muted but it has now re-emerged. The Commissioner for Agriculture in the EEC is, for instance, currently reported to favour price changes for products such as milk where chronic surpluses are an indication of excess production resources in use in the Community.[2] The other impediment is the usual absence of agreed, long-term indicative perspectives for agriculture. Few countries indeed have expressed clear ideas as to the kind of national agriculture which they wish to see evolve.

In the course of structural adjustment in developed countries, the typical forms of farm units appear to have become firmly established. Three forms predominate. The major one in the market economies is the family farm, large enough in terms of land and capital equipment to provide incomes approximately in line with non-farm occupations but small enough to require little or no regular paid help. Although there are still many family farms in Western Europe which have not yet reached an economically viable size, the trend is strongly in this direction. The second form is the large-scale corporation farm (which may still be family-owned and managed) which is becoming more frequent in 'factory-type' farming (e.g. poultry, vegetable, other field crops, and fruit and oilseed production). Their equivalent in centrally planned countries is the large-scale unit, which ranges from State farms to various forms of cooperatives and which provides the larger part of national output. (Poland is an exception in that individual farms still provide about 80 per cent of output.) Such large-scale units may include a considerable degree of vertical integration, ranging from the manufacture of production requisities to the production, processing and marketing of farm output. The third major form which has emerged from structural adjustment of agriculture in developed countries is part-time farming. This is much more significant than is generally realized. The share of national output accounted for by farms which are only one of the sources of income for their occupiers has become surprisingly large, and by the same token agriculture provides a diminishing share of the total income of farm families. Thus in Japan between a half and three-quarters of the national output comes from part-time farms and only about 30 per cent of the total income of farm households comes from farming operations. In Sweden by the early 1970s more than half the net cash income of farm families came from non-farm sources (excluding earnings from their own forestry operations) compared with 17 per cent twenty years earlier. Of the 3.4 million families in Poland linked with agriculture in 1974, one million were engaged part-time. This meant that every third adult member of farm families had two sources of income, one from his farming activities and one from other employment. The contribution to national output in other centrally planned agricultures from part-time activities concerned with small family plots and family-owned livestock is also generally very significant. It is estimated that in the USSR in 1970, private farming (which is essentially part-time) produced nearly 30 per cent of the total agricultural production of the country.[3] In the United States, where structural adjustment of agriculture has been taken further than in most countries, the share of off-farm income in the total incomes of farm families rose from a third in 1950 to two-fifths in 1960 and slightly more than a half in 1970.[4]

Although much progress has still to be made in many developed countries in the enlargement and improvement of farming units within these polarizations, it is broadly true that a large part, perhaps the largest part, of their structural adjustment within the farm sector has now been achieved. This is not so, however, in developing countries, where the process is still in its infancy and where to a large extent progress has simply meant expanding the area on which traditional semi-subsistence farming is undertaken. Some major reasons are evident. Manufacturing and other non-agricultural industries of developing countries are too small for even their rapid growth to create sufficient jobs to draw labour out of farming at a rate which would, in turn, permit worthwhile increases in the average area of land per worker in agriculture. Thus a relatively straightforward agricultural adjustment essentially based on a pulling out of labour by industry is a luxury denied to almost all developing countries. This is of course an over-simplified description of a complex situation but it remains valid even when all appropriate qualifications are made.

In these circumstances, there are very limited possibilities in many developing countries for increasing average areas of cultivated land per farm worker. Consequently, structural adjustments must be based essentially on more intensive and at the same time more productive use of labour on much the same total area of farmed land in the country. Concurrent fulfilment of these two conditions is the technical and economic problem which must be solved regardless of the political framework within which agriculture operates. The practical approach for doing this is of course increased investment per farm worker for the application of technologies and cultural practices that are not only more highly productive but also more highly labour intensive. However, the political component of this approach to agricultural adjustment is vital, i.e. land reforms to remove the huge disparity in land ownership by which a very small minority frequently controls the largest part of the farming land, with the result that individual farms are either unnecessarily large or too small. While it is true that in only a minority of countries can the total cultivated area be increased substantially and quickly, land reforms can indeed increase the average area of effectively used land per worker through reducing the inequality in land distribution, and thereby also transform incentives.

At some time in the future, the labour needs of industry and other non-agricultural sectors will allow steps to be taken in developing countries for structural adjustment in agriculture through enlarging land areas per farmer, as continues to take place in developed countries – provided society at that time still facilitates urbanization – but for most developing countries except some in Latin America and Africa this option will not be available for several decades yet.

*National agricultural adjustment*

Regardless of how well or inadequately on-farm problems are solved, agricultural adjustment poses another set of issues for governments and society to cope with at the national level. There are essentially two sets of these issues: adjustment of aggregate farm output to aggregate domestic and external demands or supply availabilities (and vice-versa), and secondly, linkages between the farm sector and national economic and social life when both are changing.

The trends and composition of farm output must be continuously adjusted to changes in total market demands and availability of supplies. Given the relative fixity of inputs, the biological nature of farm production and the long-term nature of much of farm investment, managing production trends is no easy matter. The problem is aggravated when international markets fluctuate violently. Price is the major mechanism for securing the adjustments. Virtually all countries attempt to insulate domestic prices of farm products from the full impact of changes in international markets. This is done in the interests of both producer incomes and consumer costs, the relative importance given to the two objectives varying greatly over time and amongst countries.[5] Experience indicates that maladjustments can easily take place in a regime of partly-administered prices. Producer prices which may be kept too high in the interests of improving farmers' incomes will generate surpluses. As noted earlier, dairy products in the EEC have been a chronic example and still remain so. On the other hand, consumer prices may be kept so low, particularly for staple foodstuffs, that either they give little incentive to farmers or the subsidy becomes too weighty a fiscal burden. A danger then is that price adjustments which should have taken place gradually over a certain period are introduced too abruptly. The essence of the problem is the wise use of (partly) administered prices to link trends in supply and demand. Whatever the shortcomings of administered prices, few governments are now prepared to leave supply – demand adjustments to the market price mechanism alone. Its unmodified operation is too harsh, unstable and unpredictable.

So as to have a benchmark for formulating and carrying out their efforts to keep supply – demand trends in reasonable balance, some countries have adopted objectives as to desirable degrees of self-sufficiency for food or for major commodities. Sweden, Japan and Poland are all examples of countries which have included this approach in their agricultural policy, in the case of Sweden to help to keep production from expanding too rapidly and in the case of Japan to help to plan increases in output. Poland follows a policy of approximately matching the aggregate value of agricultural exports and imports so that the maintenance of a higher level of imports is permitted only if agricultural exports are also raised. Feelings of uncertainty as to the food supplies of the future arising from the events of

1973–5 have led more countries to consider raising their national degrees of self-sufficiency, especially for basic foods.[6]

Objectives couched in these terms are undoubtedly useful in making for long-term coherence in matters pertaining to the food and agriculture of a country, but they can create problems at the international level. They are also more feasible for developed countries which have the financial capacity to pay subsidies either to increase national output or to raise the effective demand facing the national farm. There are many technical instruments for doing so, although one of the lessons which has had to be re-learned is that no country can afford for long to try to balance supply – demand in complete isolation from conditions on world markets. Developing countries without the financial strength to subsidize either producers or consumers of agricultural products, and a few of the higher-income countries particularly dependent on agricultural exports must, on the other hand, adjust themselves more quickly to the vagaries of the international market.

The influence of structural adjustment in agriculture on economic life is the second aspect at the national level. In developed countries, this role is most concerned with its effects on the cost and availability of domestic food supplies and, increasingly, on environmental and social implications of changes in agriculture. However, most affluent countries appear ready to accept whatever costs are involved in an output which at least maintains 'normal' levels of self-sufficiency. These issues of cost, environment and so on are of course also relevant in developing countries, but three other imperatives arise in connection with agricultural adjustment in the latter countries. In the course of fostering structural adjustment of agriculture, these countries must ensure that (1) domestic production meets basic needs for which imports cannot be afforded, (2) production is increased rapidly enought to 'get agriculture moving' so as to stimulate national economic growth, and (3) productive employment in agriculture is expanded, not decreased. The latter imperative has come to be seen as an ever more important task for agriculture. Urban agglomerations in developing countries are already too many, too big and have come about too soon. More employment must be provided in rural areas, and while this does not mean exclusively agriculture, that sector must provide additional employment. 'Integrated rural development', which has so far been a slogan rather than a policy, must in future become a reality. Experience in socialist economies with agro-industrial complexes and regional development schemes is also relevant to the problems faced by developing countries.

*International agricultural adjustment*

In a world of countries fully self-sufficient in both agricultural commodities and the means to produce them, agricultural adjustment would stop at national boundaries. But this of course is not the actual world. What takes place 'down on the farm' determines the supply – demand balance of a country, and this balance – or imbalance – is at the next stage the basis for international trade, to obtain additional supplies needed to meet domestic demand or to find a profitable outlet for production surplus to domestic requirements. In turn, what happens on international markets influences decisions of farmers and consumers in any country. Only about one-eighth of world agricultural output is traded internationally, but this trade is of much greater importance than might be suggested by its ratio to production. It reconciles agricultural production and demand at the national level; it permits a considerably higher standard of living by adding to the quantity and quality of diet and providing some materials essential to industry; it is a chief source of foreign earnings for many countries and it is an integral part of the system of international trade and payments of an increasingly inter-dependent world.

Trade is thus the essence of agricultural adjustment at the international level, and the conditions under which it takes place are of crucial importance to the entire adjustment process. International trade in agricultural products and requisites is characterized by many and pervasive forms of intervention. In particular, governments long ago found all kinds of ingenious ways of sharing at least part of their problem of agricultural adjustment with other countries by subsidizing exports or limiting imports when their own output was excessive or when it seemed desirable to encourage their own producers. The use of administered prices to dis-integrate national and international market mechanisms and some of the problems which arise from this were noted earlier.[7]

A crucial issue exists in agricultural adjustment at the international level because production and demand are growing at divergent rates within (broadly) developed and developing countries. Modern technology has given agriculture the capacity to expand output of food and other products relatively rapidly. However, in developed countries demand is projected to grow much more slowly than established production trends, while developing countries have not yet been able to bring about the faster growth in output to meet projected demands and to obviate the need for food imports on which they should not be spending their scarce foreign earnings. The arithmetic is simple:

Percentage of annual increases in developed and developing countries, exclusive of centrally planned economies

| | *Developed countries* | *Developing countries* |
|---|---|---|
| Food production | | |
| Latest 10 years (1965–75) | 2.4 | 2.9 |
| Food demand | | |
| Projected 1969–71 average to 1985 | 1.4 | 3.6 |

To bring trends of food production and demand more closely into line in each of the groups of countries is the adjustment challenge for agriculture faced at the world level. Balance must be achieved mainly by changing the trends of production rather than those of demand. Consumers in developed countries can afford supplies to meet their increasing demands of the future from either increased domestic production or expanded imports. However, consumers in developing countries who also need supplies fully adequate to cover the projected increases in demand if their nutrition is to be improved even modestly, cannot generally afford increased imports. It is only to their own farms that in the main they can turn for the larger supplies. Until developing countries have actually achieved a sustained improvement in production, it would however be premature to retard the upward trend in production in developed countries. Stocks need rebuilding and, as in the past twenty years, global balance can also to some extent be brought about by massive transfers of food aid to developing countries – provided it is absorbed carefully and selectively. If developing countries were successful in solving some of their production problems, global balance would then require some slowing-down in output trends in developed countries if prices were not to fall. Over the longer run, developed countries have the flexibility and resources required to bring about such a slowing-down of upward trends in production.

Finally, it should be noted that there may be unexploited scope for some adjustment through managing demand to a greater extent if the problem should be one of global pressure on food supplies. Developed countries have a large potential for economizing on supplies by reducing the quantities of grains consumed indirectly, that is via livestock feeding. The realization of this potential saving and the translating of it into additional supplies of grain actually available in deficit countries would, however, raise complex problems. In a situation of global market shortages, price elasticities in developed countries could be expected to reduce the quantities of grain used for livestock feeding, but this would not automatically make more grain available in deficit, developing countries.

The issues of international agricultural adjustment thus centre on comparative rates of national growth in production and demand. Recent years have brought increasing awareness that adjustment at this level must take the shape either of a significant increase in food output in most developing countries, so as to moderate the growth of their imports, or of a massive increase in imports on concessional terms or as food aid. There is less anxiety now than during the last food crisis of 1973–5, but in sober reality there is no reason to believe that the problem has been solved. The growth in the demands of developing countries on their agricultural sector require a rate of production growth which few have so far been able to achieve and to sustain.

Given this situation, it may seem illogical to bring up the question of the expansion of agricultural exports from developing to developed countries, but this also is a basic element in global adjustment issues. Should developing countries switch the use of agricultural resources from production for export markets to production for domestic consumption so as to do something to improve food intakes of their own people rather than trying to expand exports? Undoubtedly there must be some instances where this choice would represent the best strategy for the country, but only when alternative sources of expanding foreign earnings are present or can be brought into being. Even in such instances, increased export earnings from an expansion of manufactures or minerals or increased tourism is not likely to solve the income problems of large groups of farmers. Experience emphasizes that higher levels of domestic food consumption require higher incomes and not only increased availability of supplies. As pointed out in the case study of the Sudan, although switching from export to domestic market production would improve available supplies, higher incomes depend primarily on additional and more productive employment which can be provided only by broadly based development of the over-all economy. Relatively large imports of capital equipment and other requisites are needed for such development, and to finance these imports many developing countries depend heavily on agricultural exports. The question is not therefore a straightforward choice between using resources for export production as against production for the domestic market, but rather the proportions in which future production is to be shared between the two uses and the timing of these changes. The case study on the Sudan illustrates how an expansion in agricultural exports can be an integral part of a strategy of over-all development, of which a major objective is a permanent improvement in nutritional standards. Since that report was written, ambitious plans have been carefully drawn up for the use of substantial Arab funds to increase agricultural production both for domestic consumption and for export to other Arab countries. Subsequent Sudanese experience will be a valuable guide as to how developing countries can best be helped to realize their potential as suppliers of raw and processed agricultural products to external markets – provided those markets are willing to give assured, uninterrupted access.

Developing countries themselves place a high priority on expansion of agricultural exports to developed countries, and rightly consider that questions of assurance of access and greater stability in export earnings rank high among those requiring action. Studies of agricultural adjustment in developed countries (the studies reproduced in this volume, and particularly those listed in the Appendix) have included consideration of the feasibility of more imports being taken from developing countries. Although it was shown that the accommodation of additional supplies could raise some difficulties where they competed with domestic production or processing, the extent, pace and magnitude of change actually under way in the agriculture of developed countries for their own benefit was so great that any additional adjustment in favour of developing countries could, it appeared, be absorbed without undue strain, although in some instances adjustment assistance could properly be called for.

Within the growing acceptance of the need for some radical reshaping of the world economy, reflected by the philosophy of the new international economic order as evolved in the United Nations, international agricultural adjustment finds its place. The poor performance in recent years of agriculture in the developing countries – a decline in growth from 3.1 per cent a year in the sixties to 2.7 per cent in the first half of the seventies – is evidence enough that structural adjustment in agriculture at each of the three stages or levels must be given greater weight in future development strategies. As can be seen from various declarations and resolutions of international conferences, a broad consensus already obtains as to the major objectives for agriculture. It is, however, not surprising that wide divisions of opinion are revealed once an effort is made to translate such broad objectives into more specific and concrete policies and targets. The existence of widely differing stages of and approaches to adjustment at the national level implies formidable practical difficulties in handling international aspects. This is particularly so in view of the prevalence of national policies favouring increased food self-sufficiency. Up to the present, countries have, by and large, not been willing to enter in any substantial way into commitments regarding the impact of their policies and actions in agriculture on other countries. In other words, the international stage of agricultural adjustment in the sense of some degree of effective harmonization of national policies still remains to be achieved. Nevertheless, at least some of the preconditions now exist through a greater acceptance of the need of purposive structural changes in the world economy.

For the agricultural sector, a set of objectives and policy guidelines, including several quantitative targets, have been accepted by the member countries of the FAO as formally constituting a strategy of international agricultural adjustment. This acceptance carried only the commitment to take the guidelines into consideration, i.e. the strategy has indicative status only. However, it can be built on. The first way is through the periodic monitoring of progress in agriculture in relation to the agreed objectives and guidelines – the initial monitoring, covering approximately the period from the World Food Conference in November 1974 to mid-1977, is scheduled for presentation to the 1977 biennial FAO Conference – and, secondly, by the use of the above strategy to spell out in more detail the implications for agricultural policies of the future over-all development strategy when this is formulated in the context of the new international economic order.

In such ways can the vast structural adjustments be considered within a global framework – adjustments which must take place in world agriculture as population and incomes continue to grow and the aspirations of the under-privileged for a new deal are heeded. This will not solve the difficult problems of reconciling inconsistent national policies, of ensuring that adequate weight is given to long-term investment in agriculture and similar questions, but without such a global framework international aspects of agricultural adjustment would be even more intractable.

NOTES

1 In the study 'Agricultural Protection and Stabilization Policies: A Framework of Measurement in the Context of Agricultural Adjustment', C 75/LIM/2, FAO, Rome, 1975, Professor Josling quantified the effects of intervention measures on producer receipts and consumer expenditures in a number of countries for the period 1968–74.
2 *Financial Times*, London, 28 January 1977.
3 'USSR: A Case Study in Agricultural Adjustment', FAO, Rome, 1973, pp. 4–5.
4 The trend as quoted in D. Gale Johnson, *Farm Commodity Programs*, Evaluative Studies No. 7, Washington, D.C., May 1973, p. 15.
5 'Agricultural Protection and Stabilization Policies', op.cit.
6 See 'National self-sufficiency in food', J. P. O'Hagan, *Food Policy*, Nov 1976.
7 The use of what are called here 'administered prices' and the consequent tendency towards dis-integration of markets for agricultural products is no new event in economic affairs. See, for example, Wilhelm Röpke, *International Economic Disintegration*, Geneva, 1942.

# I Case Study of Japan

## CONTENTS

# 1 Recent Agricultural Adjustment Experience

## AGRICULTURE IN A RAPIDLY GROWING ECONOMY

Large changes have taken place in Japanese agriculture since the end of the Second World War. During the period of economic rehabilitation (1946–54) following the war, the radical land reform wrought an unprecedented structural change. In 1946 the contribution of agriculture to the national income amounted to about 30 per cent; due to the rapid expansion of other sectors, the share of agriculture had declined to about 10 per cent by 1960. The very rapid economic growth which Japan experienced during the 1960s caused the relative position of the agricultural sector to decline further. By 1971 the net domestic product at factor cost originating in agriculture had declined to 4.7 per cent of the total net domestic product. This was mainly because of the rapid growth of the manufacturing and service sectors. In fact, the gross national product of Japan had increased by nearly three times during this decade. Although the rate of growth of GNP varied from year to year it was generally well above 10 per cent per annum in real terms. In 1971 GNP of Japan amounted to US $243 billion (81,093 billion yen) and per caput national income was US $1874 (629 thousand yen).

Table 1.1 summarizes the relative position of agriculture in the Japanese economy at the beginning, in the middle and at the end of this period. The share of the agricultural labour force[1] in the total labour force fell from 27 per cent in 1960 to 14 per cent in 1971. Comparing the two series of relative shares of agriculture one can logically conclude that there is lower productivity of labour engaged in agriculture than in other sectors of the economy. As measured in terms of net national product per worker at current prices, agricultural labour productivity in 1971 was only 30 per cent of that in the manufacturing industry. The ratio of labour productivity in agriculture to that in manufacturing rose from 1960 until 1967 when it reached a peak of 39 per cent, but thereafter tended to fall. As prices of agricultural products related to prices of manufactured products continued to rise, this implies a fall in real terms (see col. 4 of Table 1.1).

Relatively lower productivity of labour engaged in agriculture as compared to other sectors of the economy is a feature not unique to

Japanese agriculture, but the extent of the disparity there appears to be very great. In fact in many developed countries there has been a similar disparity. However, there are several special features associated with

TABLE 1.1 The place of agriculture in the national economy

| | *1960* | *1965* | *1970* | *1971* |
|---|---|---|---|---|
| | | *(percentage)* | | |
| Share in net domestic product | | | | |
| Agriculture | 10.2 | 8.2 | 5.5 | 4.7 |
| Agriculture, forestry and fisheries | 14.6 | 11.1 | 7.5 | 6.5 |
| Share in labour force | | | | |
| Agriculture | 26.8 | 20.6 | 15.9 | 14.3 |
| Agriculture, forestry and fisheries | 29.7 | 23.1 | 17.1 | 15.6 |
| Productivity[a] as percentage of that in manufacturing | | | | |
| Agriculture | 28 | 35 | 31 | 30 |
| Agriculture, forestry and fisheries | 36 | 43 | 39 | 38 |
| Price indices of agricultural products as percentage of wholesale price indices of manufactured goods | 55.9 | 83.6 | 100 | 102.8 |

[a] Productivity is calculated here by dividing the net domestic product at current prices in agriculture (or manufacturing) by the number of labour force in agriculture (or manufacturing).

Japanese agriculture which have made it difficult to reconcile agricultural policy objectives designed to influence the production of farm commodities and those aimed to increase the productivity and incomes of farmers and farm workers.

First, Japan may be considered as an exceptional case among developed countries in that the government deliberately sought to meet a significant part of the increased domestic demand for food by increased imports of agricultural commodities. The declining self-sufficiency ratios in agricultural products and increasing reliance on imports of agricultural commodities are clearly indicated by Table 1.2. Though there is some difference in the coverage of food, especially in respect of fish, the data in the table give a broad picture of the situation. Thus food consumption nearly doubled during this period, while food production increased by only about 20 per cent. As a result the total volume of food imports more than quadrupled during the same period.

Secondly, the structure of Japanese agriculture presents certain characteristics which lead to difficulties in raising productivity. Agricultural land is scarce and there are a large number of small farmers with exceedingly small plots. As established historically, this pattern was suitable for labour-intensive paddy cultivation in the monsoon climate. During the last

TABLE 1.2 Indices of food consumption, production and imports, 1960–71

| | *Index of food expenditures (at 1965 prices)* | *Index of food production* | *Quantum index of food imports* |
|---|---|---|---|
| | | *(1965 = 100)* | |
| 1960 | 74 | 90 | 39 |
| 1961 | 78 | 92 | 48 |
| 1962 | 83 | 96 | 54 |
| 1963 | 88 | 94 | 68 |
| 1964 | 95 | 98 | 83 |
| 1965 | 100 | 100 | 100 |
| 1966 | 108 | 105 | 114 |
| 1967 | 117 | 113 | 119 |
| 1968 | 122 | 118 | 127 |
| 1969 | 129 | 116 | 144 |
| 1970 | 134 | 114 | 162 |
| 1971 | 141 | 108 | 177 |

decade Japan's rapid economic growth has involved a movement of labour out of agriculture into other occupations. This movement has not however been accompanied by a proportionate, or even nearly proportionate, reduction in the number of farms and increase in the size of them. Since the size of holdings is small, farmers tend to keep them for homes and for subsidiary subsistence production even when they are no longer primarily engaged in farming. High and soaring land values have encouraged farmers now engaged in non-agricultural activities to retain their land in anticipation of further increases in its market value. They have also been reluctant to lease their land to other farmers because of the legal rights that have accrued to tenants under the post-war land reform. Consequently, as will be examined later, the general trend is to increase the proportion of part-time farmers. The predominance of part-time farming suggests the extent to which the incomes of Japanese farm households are supplemented by earnings from non-agricultural activities. Although there have been cases of land consolidation, these have mainly involved the consolidation of fragmented parcels of land, and they have contributed little to the enlargement of the total size of the holdings. The enlargement of farm holdings that has occurred has been mostly the result of land reclamation and development, particularly in northern parts of the country.

## BASIC LINES OF AGRICULTURAL POLICY SINCE 1961

### *1 Basic orientation*

Rapid growth of the Japanese economy after 1955 has brought significant changes within the agricultural sector and in the relation between it and the rest of the economy. The disparity between farm and non-farm incomes tended to widen, and the incomes of people living on farms had to be raised at an unprecedented rate in order to match fast-rising urban incomes. The agricultural labour force was reduced. Demand for higher quality and a greater variety of foods increased rapidly. To cope with these changes, far-reaching modifications of agricultural policy were called for. During the early post-war years these policies had been oriented primarily towards rural democratization through the land reform, reorganization of the cooperatives, and the stimulation of food production. As a result of the sweeping land reforms absentee landownership was completely eliminated, and the majority of farmers became owners of the land that they tilled. This process, however, left the former structure of small-scale holdings practically intact. With greater incentives, new farmer – owners worked more intensively and their incomes improved. They were able to save and could afford to invest more in their farm operations. At least for a certain period of time, they were able to increase their productivity and the output from their holdings, however small. However, it now became apparent that with changing conditions the small-scale farm structure was becoming an obstacle to further increases in agricultural productivity and farm incomes.

Thus an Agricultural Basic Law was enacted in 1961. It stated that the ultimate objectives of national agricultural policy were to narrow the productivity gap between agriculture and other industries and thereby to ensure farmers living standards commensurate with those of wage-earners in other sectors of the economy. To this end it pledged the government to take necessary measures in an integrated way for the following purposes:

(a) to adjust production of agricultural commodities to the changing national demand and to pursue selective expansion of agricultural production;
(b) to improve productivity in agriculture and to increase total Japanese agricultural production;
(c) to improve agricultural structures, including enlargement of individual agricultural holdings and agglomeration of farm land; to introduce livestock and farm mechanization; to rationalize land tenure and modernization of farm management;
(d) to rationalize marketing and distribution for the promotion of food processing industries; and to expand demand for agricultural products;
(e) to stabilize prices of agricultural products and to protect agricultural

income in order to remedy unfavourable conditions of production and internal terms of trade;

(f) to rationalize production and distribution of agricultural requisites and to stabilize the prices of these;

(g) to develop the manpower potential of the rural population by training in modern methods of farm operation and management, and also training for employment in non-farm occupations;

(h) to improve rural welfare for the benefit of farmers.

Among these areas of action, the selective expansion of agricultural production and the improvement of agricultural structures represented relatively important departures from previous policies. The former meant increasing output of those agricultural products for which demand was growing and conversely checking the output of those for which demand was stagnant or declining. The latter meant, *inter alia*, the fostering of viable family farms by enlarging the size of holdings and modernizing technology and management. As well as the emphasis on establishing self-sustaining family farms, importance was attached to cooperative or group operations in agricultural production.

*2 Major policy measures taken*

Within this reorientation of agricultural policies, various new policy measures were introduced. With regard to production, particular stress was placed upon measures for promoting production of dairy and other livestock products, of fruits and vegetables, as well as for raising productivity in the growing of rice and various upland crops. In order to achieve these goals, research and extension activities were expanded and intensified. Land improvement and development projects were reinforced, with considerable emphasis on the formation of larger plots and fields and on the construction of farm roads to ensure efficient use of farm machinery. Development of permanent grassland was also emphasized.

With respect to price policies, new schemes were established to stabilize prices of pork and milk products through market intervention by the Livestock Production Development Corporation. Deficiency payments were made to dairy farmers for producing milk for manufacturing purposes and also to the producers of certain kinds of vegetables during specified seasons when market prices fell. Thus price policies covered all the major agricultural products which comprised more than two-thirds of the total sales by farmers. However, the methods and degree of intervention varied from product to product. Apart from tobacco, which is under government monopoly, intervention has been most comprehensive and strongest in the case of rice. The major part of the rice produced for sale is bought by the Government at stipulated prices or its sale supervised; the remainder is either consumed in the producing farm households or sold directly to cooperatives and wholesalers under the 'government-approved private sales scheme'.

In the field of structural policies, village-level projects for improvement of farm structure were initiated. These projects consisted mainly of land improvement and consolidation, and the installation of modern equipment and facilities for production or marketing. A legal basis was also provided for establishment of joint farm enterprises. Under a scheme called 'occupational change benefit system' allowances were made available for vocational training, and later a special scheme was introduced to make more opportunities available to members of farm families and to provide a subsidy for travelling and other expenses incurred when they were seeking jobs in distant areas. Removal compensation grants were given to those who moved to other localities to take up off-farm employment. As for land consolidation policy, the fund for long-term and low-interest loans for land acquisition was augmented and made available in an expanded scale. Attempts were made repeatedly to amend the Agricultural Land Law to facilitate the mobility of ownership and tenant farming. However, the political pressure to preserve the results of the post-war land reform was so strong that the amendment was not passed by the parliament until 1970.

Long-term loans at low-interest rates were made available on an expanded scale for various investment purposes from the Agriculture, Forestry and Fisheries Finance Corporation. In addition, the agricultural modernization loan scheme was established to enable agricultural co-operatives to provide farmers with loans (mostly medium-term) for the purchase of machinery and livestock, the construction of farm buildings and structures, and so on. The interest rates to be paid by the borrowing farmers were kept low through government subsidies.

During the first several years of the so-called 'Agricultural Policy under the Basic Law', the administration apparently placed too much emphasis on rice production. Rice had been by far the largest crop (representing almost one-half of the total agricultural output in 1960) and was grown by the great majority of farmers, both large and small. A considerable amount of investment for land improvement and enlarged research and extension activities was made, in the main, to increase rice production. The price policy for rice also played a significant role in increasing production and raising the incomes of farmers. This policy, in fact, provided a relatively easy and quick means of realizing income parity which farmers increasingly claimed. From the 1960 rice crop onwards, the official procurement price for rice was determined on the basis of the 'production cost and income compensation' formula, which allowed remuneration for family labour at a rate equivalent to the average hourly wage rate of factory workers. With a steep rise in the wage rates of manufacturing workers, the rice procurement prices to producers were also raised sharply year by year to a level, estimated by FAO in 1967, of more than twice the unit value of the world rice exports. The procurement price also exceeded the government selling price by an increasingly large margin, and resulted in mounting deficits of the Special Treasury Account for rice procurement. In 1960 about one-fourth of the total national budget for agriculture had been accounted for by price and distribution policies which, for the most part, related to rice; by 1967 the proportion had increased to nearly one-half. The agricultural policy had succeeded in producing a larger quantity of rice, but it had also resulted in accumulation of a large rice surplus in the hands of the government. Measures were then adopted to curtail rice production, and there was pressure for a renewal of structural improvement measures.

In 1970 the Agricultural Land Law was at last amended in several respects, and the existing limitations on ownership of leased land and the ceiling on rents were repealed in order to facilitate enlargement of farm holding through leasing arrangements. Agricultural cooperatives were now entitled to operate farms at the request of their members. Non-profit corporations formed to rationalize land holdings were allowed to buy, sell and sub-lease farm land for the purposes of enlarging the size of agricultural holdings, and were entitled to receive grants-in-aid from the government for their activities. On the other hand, assistance was reinforced for vocational training and occupational change for farm family members. A farmers' pension scheme was inaugurated for the purpose both of providing farmers with greater security in old age and of encouraging early retirement. Not only could the newly established Farmers' Pension Fund pay out management transfer pensions and old-age pensions in addition to the universal nation pension, but it was also authorized to buy and sell farm land owned by outgoing pensioned farmers and to pay indemnities to farmers who ceased their agricultural operations. In 1971, moreover, a law was passed to promote planned development of manufacturing and other industries in rural areas to absorb farm labour and to improve agricultural structures. These measures have certainly activated structural policies to a great extent, though it is still too early to appraise fully their results.

In 1972, out of a total amount of 1217 billion yen in the Japanese Government's total agricultural budget, 32 per cent was devoted to production policies, another 32 per cent for price, income and distribution policies, 17 per cent for adjustment of rice production, and 4 per cent for structural improvement policies.

### *3 New policy needs*

On the basis of the Agricultural Basic Law, Japan's agricultural policies have, in effect, concentrated largely on measures for promoting rice production and supporting producer prices for rice. From 1967 onwards stocks of rice in the hands of the government accumulated markedly, and by 1970 they amounted to 7 million tons. This was an unusual event, and agricultural policy was required to change its direction and to curtail rice production for the first time.

Furthermore, new dimensions were added in national agriculture as a consequence of rapid economic growth, and old problems were accentuated. These included to name a few, widespread pollution which affected marine resources and endangered the safety of fish and some livestock products; deterioration of the natural environment by disorderly development of land into residential and industrial sites; the price of land, which rose to unprecedented levels; concentration of population in industrial and city zones, which led to excessive migration from some remote rural districts, resulting in serious disruption of community life, social services and other infrastructures.

Thus there emerged new policy needs for agriculture. The environmental problem is a case in point. Agriculture has suffered water, soil and aerial pollution on one hand, and caused pollution by the use of agricultural chemicals on the other. Agriculture now has not only to prevent the use of harmful chemicals but also to contribute positively to conservation of the ecological system as well as that of the nation's land. Another important role of agriculture is the provision of green open space, which nowadays has tended to become a scarce commodity but is increasingly required for the recreation and welfare of the people. Agricultural policy thus needs to take more fully into account the role of agriculture in the advancement of the welfare of the entire population in addition to the economic efficiency of the agricultural production itself.

## INCREASED DEMAND FOR FOOD

The domestic demand for food continued to increase steadily through the 1960s and early 1970s, despite a decline in the percentage of total personal consumption expenditure that was devoted to food – from 43 per cent in 1960 to 33 per cent in 1971. During this period, in fact, total expenditures for food (at 1965 constant prices) increased by 90 per cent, or at a compound annual rate of 6 per cent. This marked increase, which was probably unparalleled among developed countries, resulted mainly from higher levels of per caput food consumption and shifts to higher-quality and more varied diets. Per caput consumption of starchy foods such as potatoes and cereals declined, while that of protein or vitamin-rich and more highly processed foods increased; more frequent dining out also contributed to the shift to a richer, more expensive diet. The rate of population growth during the period was only about 1 per cent per annum.

Empirical analysis usually demonstrates that the income elasticity of demand for food declines with improvement of per caput income. Such an analysis in Japan based on cross-section data shows that the income elasticity of demand for food in urban households declined from 0.59 in 1963 to 0.48 in 1970. However, an analysis based on time-series data of urban households in Japan covering the period 1955 to 1970 shows a slight upward tendency: 0.59 (1955–62 average), 0.61 (1955–70 average), 0.63 (1963–70 average). This apparent upward tendency is probably explained by the effect of the change in the composition of the diet over the period.

In terms of per caput consumption expenditures in urban households,[2] the expenditures for staple food (mainly rice) decreased from 1955 to 1970 by 20 per cent, while the total food expenditures increased by 1.6 times. The expenditures for dining out, luxury food (confectionery, fruits, beverages, alcoholic drinks, etc.) and side-dish food (animal protein food, vegetables, their processed food, etc.) increased by 4.1, 3.2 and 1.2 times respectively during the same period. There will be further qualitative upgrading of the diet and higher expenditures, since another feature of the evolution of Japan's domestic food demand in recent years is increasing preference for more tasty and higher-quality food, even at higher prices.

The results of the above analysis based on urban household survey data are also supported by the similar result based upon market survey data, as well as the data of national food balance sheets. According to the national Food Balance Sheets of the Ministry of Agriculture and Forestry, the average per caput daily intake of calories increased from 2290 in 1960 to 2477 in 1971, that of protein from 70 to 78 grammes, and that of fat from 29 to 54 grammes (Table 1.3). The share of starchy foods in the total calorie intake dropped from 69 per cent in 1960 to 54 per cent in 1971, while the share of animal origin in the total protein intake rose from 31 to 42 per cent.

The annual average per caput consumption of cereals declined from 150 kilogrammes in 1962 to 127 kg in 1971. This was caused by a decrease in the consumption of rice, which fell from 118 kg in 1962 to 93 kg in 1972, while other cereal consumption remained more or less unchanged. In the last eleven years the average per caput consumption of potatoes dropped by nearly half and that of pulses remained about the same, while most other commodities showed marked increases. Particularly outstanding have been the increases in consumption of livestock products and fats and oils; per caput consumption more than quadrupled in the case of meat and more than doubled in the case of eggs. The levels of consumption of dairy products and fats and oils, however, are still low compared with those in other developed countries, partly due to the traditional dietary habits in Japan. During the same period (1960–71) the per caput consumption of sugar and fruits also increased by 80 per cent and 73 per cent respectively. Even that of vegetables, which had already been at a high level, rose by a further 20 per cent.

Undoubtedly the largest single factor contributing to such marked changes in the pattern of food consumption has been the rising level of income, especially of families with relatively lower incomes. Farm families, which previously had incomes relatively lower than non-farm families but have now achieved approximate equality, are therefore gradually catching up to the food consumption standards of non-farm families in terms of both calories and quality. There have also been other contributing factors, such as the growth in the urban population, the urbanization of rural life, school lunch programmes, dissemination of nutritional knowledge assisted

TABLE 1.3 Trends of food consumption, 1960, 1965, 1970 and 1971

| | *1960* | *1965* | *1970* | *1971* | *1971 over 1960* % |
|---|---|---|---|---|---|
| Per caput intake per day: | | | | | |
| Calories | 2290 | 2408 | 2472 | 2477 | 108 |
| Share of starchy food (%) | 69 | 63 | 55 | 54 | — |
| Protein (grammes) | 70 | 74 | 76 | 78 | 111 |
| Animal protein (grammes) | 21 | 27 | 32 | 33 | 157 |
| Animal protein as percentage of total protein (%) | 31 | 36 | 41 | 42 | — |
| Fat (grammes) | 29 | 40 | 52 | 54 | 186 |
| Per caput consumption per year[a] (kg): | | | | | |
| Cereals | 150 | 145 | 129 | 127 | 85 |
| Rice | 115 | 112 | 95 | 93 | 81 |
| Wheat | 26 | 29 | 31 | 31 | 119 |
| Barleys and miscellaneous grains | 9 | 4 | 3 | 3 | 33 |
| Potatoes | 31 | 21 | 16 | 17 | 55 |
| Starch | 7 | 8 | 8 | 8 | 114 |
| Pulses | 10 | 9 | 10 | 10 | 100 |
| Vegetables | 100 | 110 | 116 | 120 | 120 |
| Fruits | 22 | 29 | 38 | 38 | 173 |
| Meat[b] | 3 | 7 | 12 | 13 | 433 |
| Eggs | 6 | 12 | 15 | 15 | 250 |
| Dairy products | 22 | 37 | 50 | 51 | 232 |
| Fish and shell fish | 28 | 29 | 32 | 33 | 118 |
| Sugar | 15 | 19 | 27 | 27 | 180 |
| Fats and oils | 4 | 7 | 10 | 10 | 250 |

a Net food supply as defined in the Food Balance Sheet.
b Excluding whale meat.

by well-developed mass media, and so on. In addition, food consumption has sometimes been stimulated by new conditions on the supply side, as demonstrated by the development of modern food-processing industries and the technological progress which has made it possible for farmers to supply almost all kinds of vegetables at any time of the year.

The index of consumer prices of food rose by about 6 per cent annually between 1963 and 1971, reflecting partly a continual rise in farm prices and partly increases in service charges for processing and distribution. The rate of increase in consumer food prices had been more or less comparable to that of general consumer prices; however, from 1967 onwards the rate of increase in food prices surpassed the general consumers price increase. Between 37 and 46 per cent of the annual change in the general consumers price index was due to the change in food items during 1969 to 1971. Furthermore, out of this large contribution of food items to the change in the general consumer prices, nearly 80 to 90 per cent was contributed by non-staple food items such as fresh vegetables, fresh fish and processed food products. The government has therefore increasingly made efforts to stabilize the consumer prices of these items through accelerated import and through the rationalization of domestic marketing and distribution systems.

It is indeed important to note (as shown in Table 1.4) that the gap between the total national expenditure on food and the value of domestic food production has increased continuously year by year. The difference between the two consists, of course, in the cost of imported food items and margins charged for distribution and processing food before it reaches the final consumer.

## DEVELOPMENT IN AGRICULTURAL PRODUCTION

### *1 Changes in production patterns*

In 1967 the production of rice reached a level well beyond the domestic demand. Production in subsequent years resulted in the accumulation of stocks of rice by the Japanese Government which in turn led to the introduction of an unprecedented policy to curtail rice production.

The production of vegetables, though fluctuating because of weather conditions, has tended to increase. An important factor in the increase has been the widespread introduction of vinyl or glass hothouses which, coupled with sophisticated cultural practices, has enabled farmers to grow a wide range of vegetables throughout the year, as referred to earlier. Total fruit production has increased by nearly three-fourths in the past eleven years. In particular, mandarin orchards were almost tripled in area between 1960 and 1972, with the output tending to glut the market recently.

Livestock production in 1971 was also nearly three times as large as in 1960. The increase was especially large in pigment, broilers and eggs, followed by milk. The numbers of livestock and poultry (Table 1.5) show similar trends. In the last twelve years the number of pigs and poultry more than tripled, and that of dairy cattle more than doubled. The number of beef cattle decreased because of low profitability as the result of high costs

TABLE 1.4 Domestic agricultural output of food and national expenditure for food

| | *1963* | *1964* | *1965* | *1966* | *1967* (*100 million yen*) | *1968* | *1969* | *1970* | *1971* |
|---|---|---|---|---|---|---|---|---|---|
| Domestic agricultural output[a] | | | | | | | | | |
| Rice | 11,053 | 12,324 | 13,339 | 14,957 | 18,536 | 19,623 | 19,139 | 17,352 | 15,162 |
| Wheat and barley | 341 | 837 | 941 | 869 | 896 | 980 | 752 | 473 | 456 |
| Other grains | 45 | 28 | 30 | 32 | 34 | 28 | 26 | 21 | 19 |
| Beans | 516 | 411 | 521 | 468 | 482 | 514 | 519 | 548 | 536 |
| Tubers | 847 | 715 | 793 | 852 | 776 | 678 | 666 | 780 | 677 |
| Vegetables | 2542 | 3146 | 3577 | 4210 | 4913 | 4849 | 6177 | 7012 | 6243 |
| Fruits | 1685 | 1787 | 2090 | 2420 | 2521 | 2655 | 3546 | 3945 | 3713 |
| Others | 1029 | 1262 | 1370 | 1548 | 1825 | 2116 | 2333 | 2761 | 2961 |
| Livestock products | 4735 | 5130 | 6054 | 7011 | 7606 | 8670 | 9770 | 10,311 | 11,360 |
| Total (T) | 22,793 | 25,640 | 28,715 | 32,367 | 37,589 | 40,113 | 42,928 | 43,203 | 41,127 |
| National expenditure on food (E) | 55,344 | 62,194 | 70,147 | 78,457 | 88,921 | 99,352 | 113,039 | 127,940 | 141,994 |
| (T)/(E) percentage | 41.2 | 41.2 | 40.9 | 41.3 | 42.3 | 40.4 | 38.0 | 33.8 | 29.0 |

[a] Domestic agricultural output excluding tobacco, cocoon, and other non-food products.

of calf breeding and cattle raising. Formerly beef production depended primarily on the fattening and slaughtering of aged draft cattle. With the disappearance of these as they have been replaced by machine power,

TABLE 1.5 Area planted and livestock population, 1960, 1965, 1970 and 1971

| | *1960* | *1965* | *1970* | *1971 (or 1972)*[a] | *1971 (or 1972) over 1960* (%) |
|---|---|---|---|---|---|
| Area planted (1000 hectares) | | | | | |
| Rice | 3308 | 3255 | 2923 | 2685 | 81 |
| Wheat, barley, rye and oats | 1520 | 961 | 483 | 360 | 24 |
| Miscellaneous grains | 144 | 84 | 37 | 34 | 24 |
| Sweet potatoes | 330 | 257 | 129 | 107 | 32 |
| Irish potatoes | 197 | 202 | 151 | 148 | 75 |
| Pulses | 642 | 485 | 338 | 336 | 52 |
| Fruits | 254 | 356 | 416 | 422 | 166 |
| Vegetables | 615 | 692 | 688 | 690 | 112 |
| Industrial crops | 447 | 365 | 257 | 247 | 55 |
| Mulberry | 166 | 164 | 163 | 166 | 100 |
| Forage and green manure crops | 506 | 611 | 736 | 801 | 158 |
| Total | 8129 | 7430 | 6311 | 6002 | 74 |
| Cultivated area | 6071 | 6004 | 5796 | 5741 | 95 |
| Multiple cropping ratio (%) | 134 | 124 | 109 | 105 | — |
| Number of livestock and poultry[b] | | | | | |
| Dairy cattle ('000s) | 824 | 1289 | 1804 | 1819 | 221 |
| Beef cattle ('000s) | 2340 | 1886 | 1789 | 1749 | 75 |
| Pigs ('000s) | 1918 | 3976 | 6335 | 6985 | 364 |
| Hens (millions) | 54.6 | 120.2 | 169.8 | 164.0 | 300 |
| Broilers (millions) | — | 18.3 | 53.7 | 67.9 | 371[c] |

[a] Livestock and poultry refer to 1972.
[b] As of 1 February each year.
[c] 1971 over 1965.

cattle raising for meat production has not been particularly remunerative, despite relatively high beef prices. The comparatively long time required for raising cattle plus the combination of limited areas of grassland, high land values and generally high costs for feedstuffs has made production of beef-type cattle relatively unattractive to Japanese farmers. Beef production from aged dairy cows has increased, and also there has been a tendency to increase the number of dairy-type steers fattened for beef production. However, the dairy-type fat cattle usually bring lower prices than beef-type animals because of consumer preferences. Apart from beef cattle, the larger herds of livestock and flocks of poultry have increased their share in the total number of livestock and poultry kept. In 1972, for instance, slightly less than 7 per cent of the pig raisers had more than fifty pigs, but these larger herds accounted for about half of the total pig population. Likewise, less than one per cent of the poultry farms kept more than 3000 hens, but these larger flocks accounted for well over one-half of the total number.

On the other hand sericulture, which requires highly labour-intensive activities, remained almost stagnant despite the brisk demand for raw silk. Similarly, notwithstanding a strong and growing demand, ordinary upland crops including wheat, barley, miscellaneous grains, potatoes and pulses showed a continuing decrease in both acreage and production. The combined area planted to these crops fell to 985,000 hectares in 1971, as against 2,833,000 hectares in 1960.

Consequently there occured a big change in the percentage shares of different commodities in the total value of agricultural output. In addition to a reduced share for ordinary upland crops,[3] the share of rice declined from 48 per cent in 1960 to 35 per cent in 1971, but nevertheless remained the largest. By contrast, fruits and vegetables increased their combined share from 14 per cent in 1960 to 23 per cent in 1971; and similarly livestock products from 15 to 26 per cent.

Among the forces that caused these changes in production patterns, the most influential was apparently the farmers' increasing awareness of the comparative gross and net incomes that could be realized from the various crops, and of opportunities for alternative employment in other industries. Obviously these non-farm employment opportunities have increased as a general shortage of labour developed in other sectors of the economy and wages consequently rose. However most farmers maintained their rice cultivation, partly because its seasonality and mechanization enabled them to combine rice growing with off-farm employment for most of the year. Another reason was the highly profitable nature of rice cultivation due to the government support for the price. Although they abandoned the cultivation of wheat, barley, sweet potatoes, soybeans, and rapeseeds for more profitable crops (such as certain kinds of vegetables) or off-farm work, farmers generally continued to grow rice. Even the producer price policies of the government for these other crops were unable to assure adequate returns to the growers. In a sense it may safely be said that those crops lost in competition with imported supplies. For livestock products like pigs and eggs, profitability was not very high for the average producers but was much higher for large holdings.

The selective expansion of agricultural production has by the early 1970s been only partially achieved, however. There has been notable development in capital-intensive and less land-using lines of production because economies of scale can relatively easily be realized. On the other hand, such lines of production as upland farming and cattle raising have faced difficulties in expanding production on account of serious constraints on size of farms. As farmers have become increasingly income-conscious, they have become more attracted to off-farm employment and to rather simplified cropping patterns, with abandonment of the cultivation of less profitable crops, notably winter crops. In fact, the total planted area in 1971 was only 6 million hectares, that is, over 2 million hectares less than in 1960, and the multiple-cropping index fell to 105 per cent in 1971 as compared with 134 per cent in 1960. The spectacular expansion of livestock industry has in large part depended upon imported feeds. In fact, the 'cheap feed' policy was designed to meet the needs of the livestock industry; still in its early stage, implementation of this policy has so far relied largely on the use of imported feedstuffs.

*2 Rice surpluses and production adjustment*

Historically, rice farming has always been of overwhelming importance and, as it were, represented Japanese agriculture itself. Even though its relative importance has recently been declining, rice remains the most important agricultural commodity produced in Japan. In as late as 1971, when strenuous efforts had already been started to curtail rice production, 84 per cent of the Japanese farms grew rice, and 70 per cent of them produced it for sale. Traditionally, rice was the staple food for the people, and it still remains the largest single item in their cereal consumption (e.g. 73 per cent in 1971) and the greatest source of their calorie intake (36 per cent).

Since 1942 the marketing and distribution of rice had been under complete government control with various modifications notably in the last decade. The Food Control Special Account currently buys at fixed prices the great bulk of the rice marketed and sells it at fixed prices to the wholesalers. During and immediately after the Second World War when food supplies were seriously short, the price control measures served to secure the rice needed and to distribute it at reasonable prices. In the period under review, however the basic economic conditions have changed, and this system has functioned increasingly to raise prices received by Japanese rice producers while sheltering consumers from the full extent of these higher prices. As mentioned earlier, the official prices of rice to the producer were raised sharply year after year, in order to compensate for increasing costs of production and to make the farmers' revenue comparable to the increasing wages in manufacturing industries. Thus the prices paid to producers exceeded the sales prices by a widening margin, resulting in an increasingly heavy deficit government account.

During the early 1960s 12 to 13 million tons of rice per year were consistently produced, until 1967 when the level of production reached 14 million tons (in terms of brown rice). This level was sustained in the following two years. The increase in rice production was due partly to some additions to the area planted to rice in northern parts of the country, but mainly to higher yields per hectare all over the country. There is no doubt that government assrurance of a market for an unlimited quantity of rice at the supported prices, and the strong security against crop failures provided by the crop insurance scheme,[4] guaranteed high profitability for rice cultivation, which constituted the main incentive for farmers to increase production. Other contributing factors included the use of higher-yielding varieties, increased application of fertilizers and agricultural chemicals, introduction of modern farm machinery, extension and improvement of irrigation facilities, and popularization of more sophisticated yet standardized cultural practices.

After reaching a post-war peak of 118 kilogrammes in 1962, per caput consumption of rice (milled basis) per year continued to decrease to 93 kilogrammes in 1971. The decrease has taken place not only for urban families but also for farm families, who, however, still consume twice as much per caput as the former. Thus, the total internal demand for rice fell to 11.9 million tons in terms of brown rice in 1971, or some 1.5 million tons less than in the peak year of 1963 (see Table 1.6).

TABLE 1.6 Supply, demand and government stocks of rice

| *Fiscal year* | *Supply* | | *Internal demand* | *Government carry over stocks at the end of October* | |
|---|---|---|---|---|---|
| | *Production* | *Import* | | *Total* | *Of which indigenous old crop rice* |
| | *1000 tons (brown rice equivalent)* | | | | |
| 1960 | 12,858 | 219 | 12,618 | 4215 | 440 |
| 1961 | 12,419 | 77 | 13,062 | 3989 | 501 |
| 1962 | 13,009 | 182 | 13,315 | 4142 | 95 |
| 1963 | 12,812 | 239 | 13,410 | 3216 | 17 |
| 1964 | 12,584 | 502 | 13,361 | 2808 | 14 |
| 1965 | 12,409 | 1052 | 12,993 | 3448 | 52 |
| 1966 | 12,745 | 679 | 12,503 | 3725 | 205 |
| 1967 | 14,453 | 364 | 12,483 | 6442 | 644 |
| 1968 | 14,449 | 265 | 12,251 | 7735 | 2975 |
| 1969 | 14,003 | 48 | 11,965 | 9355 | 5533 |
| 1970 | 12,689 | 15 | 11,948 | 10,470 | 7202 |
| 1971 | 10,887 | 10 | 11,859 | 7747 | 5891 |
| 1972 | 11,897 | 5 | . . . | 5750 | 3074 |

Self-sufficiency in rice production had been a long-pursued aim of Japanese government policies. The expanded production coupled with decreased consumption brought a change in the rice balance and, for the first time in the post-war period, production exceeded consumption. Consequently, the government accumulated unsold stocks of old crop rice, which totalled 7.2 million tons at the end of October 1970 as compared with 0.6 million tons three years earlier. This further added to the already large deficit of the Food Control Special Account.

The government attempted to dispose of the accumulated stocks by both exports (mostly on a deferred payment basis) and concessional domestic sales for livestock feeds and industrial uses. However, these measures did not solve the problem and the government incurred additional losses on such surplus disposal arrangements. After a small-scale trial in the previous year, the government therefore embarked upon the curtailment of rice production for the 1970 crop by providing a direct payment to cooperating farmers. From the 1971 crop onwards, rice production adjustment was to be operative for five years and a higher rate of compensation payment was set for the diversion of paddy fields to other crops than for the mere suspension of rice cultivation.[5]

As a consequence, a total of 540,000 hectares of paddy fields in 1971 and 563,000 hectares in 1972 were withdrawn from rice cultivation. A little less than one-half of the area withdrawn was diverted to grass, vegetables and other crops in 1972, with the rest lying fallow. Moreover, to discourage rice production, the government buying price of rice for the 1969 and 1970 crops was kept constant at the 1968 level of a little over 20,000 yen per 150 kilogrammes of brown rice. For the 1971 crop, however, it was raised somewhat to allow a premium for quality rice. For the 1972 crop, in view of the increased production cost since 1968, the government purchase price from the producers was raised by 3 per cent. As a result of these measures for curtailment of rice production, the government-held stocks of old crop rice fell to a little over 3 million tons as at the end of October 1972. It is expected that surplus rice stocks will almost disappear in the course of 1973 – much earlier than previously anticipated.

Because rice is a crop of special significance for Japanese agriculture its over-production and the resulting production curtailment measures have had an adverse effect upon farm management and the farm economy in general, which has depressed the farmers' enthusiasm for successful farming and has disturbed agricultural development as a whole.

## FOREIGN TRADE IN AGRICULTURAL PRODUCTS

### *1 Agricultural exports*

In 1971 agricultural exports from Japan totalled US $390 million as against US $167m in 1960. However, their share in the total exports from

Japan declined from 4 per cent to less than 2 per cent during the same period. Export of surplus rice (worth US $134m) may well be regarded as a temporary phenomenon. Excluding these, agricultural exports in 1971 were valued at only US $256m.

Among major exports of commodities of agricultural origin, tinned and fresh mandarins (worth US $48m) ranked first in 1971, followed by dried mushrooms, mono-sodium glutamate, milk powder, and a variety of minor items.

*2 Expansion of agricultural imports*

As already indicated, the strong increase in domestic demand for food and feed outpaced the growth of domestic production. The government followed a policy of increasing agricultural imports while still maintaining controls over a number of important items. The result was a striking expansion of agricultural imports into Japan during the last decade. In 1971 agricultural imports totalled US $4448m[6] representing 23 per cent of Japan's total imports), two and a half times the 1960 value of US $1729m. The compound rate of increase during this period was about 9 per cent per annum, which was probably the highest sustained growth rate of imports of agricultural products during the decade in any large developed country. Thus Japan has become a leading importer of many of the agricultural commodities traded in the world market. In quantitative terms, Japan imported in 1970 approximately 8 per cent of the world imports of wheat, 22 per cent of maize, 61 per cent of miscellaneous grains (including grain sorghum and buckwheat), 27 per cent of soybeans, and 15 per cent of raw sugar.[7]

As shown in Table 1.7, the value of imports of cotton, wool and natural rubber, which had altogether represented 49 per cent of the total agricultural imports in 1960, did not increase substantially, and in the case of natural rubber even decreased. Strong competition from synthetic substitutes was a major reason. The combined share of these agricultural raw materials in Japan's total agricultural imports dropped to 20 per cent in 1971. In sharp contrast, nearly all other commodity groups showed enormous increases in imports between 1960 and 1971.

Table 1.8 shows imports of major agricultural commodities in 1971 compared with 1960, 1965 and 1970. The import of feed grains increased tremendously, with annual imports of maize (for feeds)[8] and grain sorghum together amounting to 7 and 8 million tons respectively in recent years. In addition increased imports of barley and wheat were in no small measure directed to feed uses. Among the oilseeds, soybeans remained the major item, with imports rising to more than 3 million tons annually. Imports of most other oilseeds also showed large increases. Particularly notable is the increased import of rapeseeds which, formerly, had been a typical, indigenous oilseed. There were also sharp increases in the imports of some tropical products. The import of bananas reached almost one

TABLE 1.7 Agricultural imports into Japan by commodity groups, 1960, 1965, 1970 and 1971

| | Value at current prices | | | | Index of value | Percentage distribution | |
|---|---|---|---|---|---|---|---|
| | *1960* | *1965* (US$ million) | *1970* | *1971* | *1971* (1960 = 100) | *1960* | *1971* |
| Crops and crop products | 750.8 | 1693.0 | 2714.8 | 2912.7 | 388 | 43.4 | 65.5 |
| Cereals and meals (including processed) | 291.5 | 767.1 | 1046.1 | 1095.2 | 376 | 16.9 | 24.6 |
| Fruits (fresh and processed) | 20.9 | 108.0 | 236.5 | 256.1 | 1225 | 1.2 | 5.8 |
| Vegetables (fresh and processed) | 10.7 | 37.6 | 51.4 | 59.1 | 552 | 0.6 | 1.3 |
| Raw sugar and sugar products | 122.1 | 183.2 | 339.3 | 376.9 | 309 | 7.1 | 8.5 |
| Confectionary, coffee, cocoa and tea | 21.7 | 61.2 | 146.6 | 135.5 | 624 | 1.3 | 3.0 |
| Other processed foods and drinks | 6.1 | 14.3 | 44.2 | 53.1 | 870 | 0.4 | 1.2 |
| Oilseeds and vegetable oils | 193.2 | 354.1 | 593.1 | 641.3 | 332 | 11.2 | 14.4 |
| Tobacco (including processed) | 15.7 | 49.6 | 73.8 | 100.0 | 637 | 0.9 | 2.2 |
| Other crop products | 69.3 | 117.8 | 176.0 | 174.7 | 252 | 4.0 | 3.9 |
| Livestock and livestock products | 131.9 | 237.9 | 448.4 | 508.6 | 386 | 7.6 | 11.4 |
| Animals | 2.1 | 7.0 | 20.5 | 25.2 | 1200 | 0.1 | 0.6 |
| Meat (fresh and processed) | 13.8 | 41.5 | 132.0 | 190.9 | 1383 | 0.8 | 4.3 |
| Dairy products and eggs | 21.4 | 31.3 | 65.5 | 74.4 | 348 | 1.2 | 1.7 |
| Animal fats | 33.3 | 32.5 | 60.0 | 61.2 | 184 | 1.9 | 1.4 |
| Hide and skin | 55.7 | 77.6 | 120.4 | 106.3 | 191 | 3.2 | 2.4 |
| Others | 5.5 | 28.1 | 50.1 | 50.5 | 918 | 0.3 | 1.1 |
| Raw silk | 0.9 | 9.1 | 84.3 | 122.7 | 13,633 | ... | 2.8 |
| Cotton, wool and natural rubber | 845.1 | 888.1 | 950.4 | 903.6 | 107 | 48.9 | 20.3 |
| Cotton | 431.4 | 441.8 | 470.7 | 515.7 | 119 | 25.0 | 11.6 |
| Wool | 270.7 | 345.0 | 350.3 | 276.5 | 102 | 15.6 | 6.2 |
| Natural rubber | 143.0 | 101.3 | 129.4 | 111.4 | 78 | 8.3 | 2.5 |
| Total | 1728.7 | 2828.1 | 4198.0 | 4448.1 | 257 | 100 | 100 |

TABLE 1.8 Imports into Japan of major agricultural commodities, 1960, 1965, 1970 and 1971

| | *1960* | | *1965* | | *1970* | | *1971* | | *1971 index* | |
|---|---|---|---|---|---|---|---|---|---|---|
| | *'000 tons* | *US \$m* | *'000 tons* | *US \$m* | *'000 tons* | *US \$m* | *'000 tons* | *US \$m* | *Quantity (1960=100)* | *Value* |
| Wheat | 2678 | 176.9 | 3645 | 251.1 | 4685 | 318.4 | 4872 | 344.2 | 182 | 195 |
| Barley | . . . | . . . | 635 | 40.9 | 767 | 41.6 | 865 | 58.7 | 136[a] | 144[a] |
| Maize (feeds) | 1306 | 78.3 | 2946 | 197.5 | 4383 | 294.0 | 3401 | 244.3 | 260 | 312 |
| Maize (other uses) | 47 | 2.8 | 487 | 34.0 | 1635 | 112.9 | 1608 | 118.4 | 3417 | 4229 |
| Grain sorghum | — | — | 1425 | 83.8 | 3773 | 230.4 | 3798 | 252.4 | 267[a] | 301[a] |
| Bananas | 42 | 6.6 | 358 | 60.6 | 844 | 144.2 | 989 | 139.3 | 2355 | 2112 |
| Raw sugar | 1264 | 106.0 | 1693 | 149.3 | 2596 | 282.7 | 2495 | 316.0 | 197 | 298 |
| Coffee beans | 11 | 8.6 | 19 | 13.9 | 81 | 55.9 | 68 | 42.5 | 618 | 494 |
| Cocoa beans | 10 | 6.4 | 25 | 10.2 | 55 | 26.0 | 39 | 23.4 | 390 | 366 |
| Soybeans | 1128 | 107.4 | 1847 | 225.7 | 3244 | 365.8 | 3212 | 421.5 | 285 | 392 |
| Rapeseeds | 51 | 6.6 | 101 | 13.1 | 336 | 41.4 | 407 | 55.5 | 798 | 841 |
| Cottonseed | 71 | 6.4 | 217 | 17.1 | 297 | 24.2 | 248 | 25.0 | 349 | 391 |
| Copra | 85 | 17.1 | 94 | 19.0 | 127 | 26.8 | 122 | 22.5 | 144 | 132 |
| Meat | 38 | 13.6 | 83 | 38.9 | 204 | 124.1 | 263 | 182.1 | 692 | 1339 |
| Cheese curd | 1 | 1.2 | 10 | 6.4 | 34 | 20.0 | 36 | 22.5 | 3600 | 1875 |
| Tallow | 158 | 25.3 | 191 | 41.7 | 252 | 53.4 | 257 | 53.7 | 163 | 212 |
| Cattle hide | 81 | 28.3 | 132 | 39.7 | 211 | 76.0 | 192 | 67.3 | 237 | 238 |
| Raw silk | — | — | . . . | 3.5 | 4 | 69.1 | 6 | 106.7 | . . . | 3049[a] |
| Jute[b] | 54 | 12.1 | 94 | 20.4 | 104 | 19.0 | 73 | 15.2 | 135 | 126 |
| Cotton[c] | 728 | 431.4 | 702 | 441.8 | 768 | 470.7 | 754 | 515.7 | 104 | 119 |
| Wool[d] | 193 | 270.7 | 242 | 345.0 | 301 | 350.3 | 296 | 276.4 | 153 | 102 |
| Natural[d] rubber[e] | 188 | 143.0 | 222 | 101.3 | 308 | 129.4 | 335 | 111.4 | 177 | 78 |

— indicates nil.
. . . indicates negligible amount.
[a] 1965=100.
[b] Including jute cuttings.
[c] For quantity, raw cotton other than linters.
[d] For quantity, greasy wool.
[e] For quantity, raw natural rubber and latex.

million tons in 1971 as against only 42,000 tons in 1960. For coffee beans and cocoa beans, the quantity of imports in 1971 was also several times larger than in 1960. Among livestock products, there was a spectacular rise in the import of cheese curd, and meat imports increased by seven times. With regard to raw silk, which used to be a staple Japanese export, Japan has since 1966 become a net importer with rapidly growing imports, which exceeded US $100m in 1971.

As indicated in Table 1.9, in 1971 Japan imported 60 per cent of her total agricultural imports from the developed countries (mostly North America and Oceania), 36 per cent from developing countries and 4 per cent from countries with centrally planned economies. Compared with 1960, the share of the developed countries was slightly higher. The United States remained the largest single source, supplying about 31 per cent of Japan's total agricultural imports.

TABLE 1.9 Percentage distribution of agricultural imports into Japan by sources, 1960 and 1971

| | *1960* | | *1971* | |
|---|---|---|---|---|
| Developed market economies | | 57 | | 60 |
| of which: North America | 40 | | 37 | |
| Oceania | 17 | | 16 | |
| Developing market economies | | 42 | | 36 |
| Asia and the Far East | | 23 | | 15 |
| Africa | | 5 | | 6 |
| Latin America | | 14 | | 15 |
| Centrally planned economies | | 1 | | 4 |
| World total | | 100 | | 100 |

The developing countries's share in Japanese imports of agricultural commodities fell from 42 per cent in 1960 to 36 per cent in 1971. This is because of the falling share of Asia and the Far East, due partly to reduced availabilities of some of this region's principal export commodities, as well as to the change of Japan's position from a rice importer to self-sufficiency. Although Japan's imports from Asia and the Far East of bananas, maize, molasses, pulses, tinned pineapple and tobacco increased notably, those of rice fell drastically and those of sugar dropped by almost half during this period. Total imports of copra from the region stagnated, although those from Pacific islands materially increased. Imports of kapok seeds also remained stagnant. Although Japanese imports of castor beans and palm oil expanded substantially, the percentage increases were much lower than those for imports from other regions of major oilseeds, including soybeans, rapeseeds, cottonseed, flaxseed, safflower seed, etc. Moreover, imports of natural rubber and fibres from Asia and the Far East also tended to stagnate. Despite a quantitative increase, the value of imports of natural rubber declined owing to a sharp drop of unit value. Imports of jute increased only by about a quarter, while those of cotton and coir remained practically unchanged and those of abaca declined to less than half. However, imports of cotton products and jute products from this region increased remarkably between 1960 and 1971.[9] During the same period on the other hand, there was a marked increase in imports of sugar, bananas, raw cotton, coffee beans and meat from Latin America, and in imports of raw cotton, sugar, cocoa, coffee beans and cottonseed from Africa.

*3 Trade policy, agricultural protection and self-sufficiency*

Since 1961 the Japanese Government has been following a step-by-step liberalization of imports. Agricultural products, including forest and fishery products, subject to quantitative import restrictions numbered 80 items[10] in April 1963. At the end of September 1969 they still numbered 73 items, but thereafter liberalization was accelerated. By the end of April 1972 the number of such items under import quota had been reduced to 24, of which 20 belonged to agricultural products in a narrow sense. In addition, import quotas for these products have generally been expanded rather substantially. Simultaneously, tariffs for many agricultural products have been reduced under the Kennedy Round or in other ways. For certain items, tariff reduction or exemption from import duty has been implemented with a view to contributing to stabilization of consumer prices, whose rising trend has become an important national issue in recent years.

The remaining import quota items (mid-1973) include beef, some milk products, rice flour and wheat flour, malt, tangerine oranges, fruit juice, tomato ketchup, glucose, starch, certain pulses. Some of the quotas relate to products for which demand is growing and which are expected to play an important role in domestic agricultural production in the future. Others apply to a group of products which are of vital importance to farmers and local communities in remote, often depressed, areas. It is roughly estimated that those agricultural products which in mid-1973 were protected by quantitative import restrictions and by state trading (which covers four items: rice, wheat, barley and certain milk products) represented altogether about 53 per cent of the 1971 agricultural output, as compared with 77 per cent in 1963.[11] In 1971, the imports of the products under quotas and state trading represented some 20 per cent of Japan's total agricultural imports.

There is little doubt that government actions directed towards trade liberalization and tariff reduction have stimulated, at least in part, a marked expansion of agricultural imports. Consequently the self-sufficiency ratio has, for nearly all agricultural commodities except rice, tended to fall. As can be seen in Table 1.10, soybeans and wheat showed the most remarkable reduction in self-sufficiency, falling to only 4 and 8 per cent respectively in 1971. The ratios for barleys and fruits also decreased to a considerable extent. The ratio of self-sufficiency in concentrated feedstuff declined sharply from 67 per cent in 1960 to 33 per

TABLE 1.10 Trend of self-sufficiency ratio for food in Japan, 1960, 1965, 1970 and 1971

| | *1960* | *1965* | *1970* | *1971* |
|---|---|---|---|---|
| | | *(percentage)* | | |
| Food of agricultural origin[a] | 90 | 81 | 76 | 72 |
| By commodities[b] | | | | |
| Rice | 102 | 95 | 106 | 92 |
| Wheat | 39 | 28 | 9 | 8 |
| Barley | 107 | 73 | 34 | 29 |
| Soybeans | 28 | 11 | 4 | 4 |
| Fruits | 100 | 90 | 84 | 82 |
| Meat[c] | 91 | 89 | 88 | 83 |
| Eggs | 101 | 100 | 97 | 98 |
| Dairy products | 89 | 86 | 89 | 88 |
| Sugar | 18 | 30 | 23 | 19 |
| Concentrated feedstuff[d] | 67 | 44 | 33 | 40 |

[a] The ratio is obtained by dividing the value of domestic production by the value of available supply for domestic consumption. Both values are assessed at the 1965 wholesale prices. Feed used including all imported feedstuff valued at the wholesale prices is deducted from domestic production in order to avoid double counting.

[b] The ratios are obtained by dividing the quantity of domestic production by the quantity of available supply for domestic consumption.

c Exculuding whale meat. [d] The ratio is obtained by dividing the quantity of indigenous concentrated feedstuff by the quantity of total supply of concentrated feedstuff. Quantities are measured in terms of total digestible nutrients.

cent in 1970, though in the following year it temporarily rose to 40 per cent on account of large quantities of surplus rice used for feed. The ratios for livestock products such as meat, eggs and dairy products have not declined very much during the same period. The aggregate self-sufficiency ratio for food of agricultural origin, as measured in terms of value (calculated by using 1965 wholesale prices) fell from 90 per cent in 1960 to 76 per cent in 1970 and, partly reflecting crop failures owing to bad weather, further to 72 per cent in 1971.[12] From the viewpoint of land use in a country, a self-sufficiency ratio measured in terms of so-called 'original calories'[13] might indicate better the contribution of production from domestic land resources to the total demand. According to this type of measure, Japanese self-sufficiency in 1971 was only 53 per cent as compared to 78 per cent in 1960.

The international competitive position of Japanese agriculture may, for some products, be considered to be relatively weak because of the very large proportion of extremely small-scale production operations. However, for some other products, such as pigmeat, broilers and eggs, a large part of commercial production is accounted for by relatively large-scale producers. The recent revaluation of the yen has also tended to reduce the international competitive position of Japanese agricultural products.[14] Thus, some governmental support for Japanese agriculture will be likely to continue in the foreseeable future. This, however, may create more acute problems of rising food prices in domestic markets unless politics and measures are carefully selected.

## CHANGES IN AGRICULTURAL STRUCTURE

### *1 Agricultural land*

In 1970 agricultural land totalled 6.3 million hectares or only 17 per cent of the total Japanese land area. Forest land occupied an additional 66 per cent. The low percentage of land used for agricultural purposes is due mainly to the topography. However, it also partly reflects underdevelopment of grassland for grazing, because dairy farming and beef-cattle rearing were launched on a substantial scale only after the Second World War. Population density per square kilometre of agricultural land in 1970 was 1642 persons, one of the highest in the world. More than 90 per cent of agricultural land, or 5.8 million hectares, were devoted to paddy fields (3.4m hectares), upland fields (1.3m hectares), land under permanent crops (0.6m hectares), and temporary grass (0.3m hectares). The area of cultivated land has tended to decrease slightly as additions to it from land development and reclamation have been more than offset by withdrawals for industrial, residential and road sites, forest and other uses.

Accelerated industrialization and urbanization and extension of the highway network in the last decade have caused land prices to rise steeply. Thus prices of agricultural land, which had already been high in comparison with those in other countries, soared further not only in the areas adjacent to urban centres but almost everywhere in the country. In rice-producing areas rising prices of paddy land can be attributed, at least in part, to the successive increase in rice prices until around 1968. Thereafter, however, land prices have still continued to rise rapidly. According to a regular survey by the National Chamber of Agriculture, the national average price of middle-grade paddy land per one-tenth of a hectare reached 1,247,000 yen in 1971 or more than six times the 1960 level; this represents a compound annual rate of increase of 18.1 per cent. In fact, since 1967, the price of paddy land has increased more than 20 per cent annually. Such a rise in the price of land, which has surpassed by far the rate of increase in prices of agricultural products or in the cost of living, has hindered transactions in farm land. Farmers tend increasingly to retain their land as valuable property rather than as a factor of production.

*2 Farm production and agricultural labour force*

Increased employment opportunities in non-agricultural sectors induced members of farm families to leave farm employment in such numbers that the farm population decreased from over 34 million (or 36 per cent of the total population) in 1960 to 25 million (or 24 per cent) in 1972. The agricultural labour force, which had since the early days of the Meiji era remained almost constant at the level of 14 – 15 million, started to decrease in about 1955. Given the generally prevailing employment practices of life-long employment and promotion by seniority which was deeply rooted in the Japanese social and economic system, new non-agricultural employment was sought mainly by the younger generation, especially those newly graduated from high schools. In 1972 only 22,000 young farm people of both sexes found work in agriculture when they left school, as compared with 74,000 in 1963. In addition, the relatively young who had already become engaged in farming also found non-farm jobs, and even older ones followed in increasing numbers, particularly after adoption of the policy of curtailment of rice production. Many of them however continue to live in the farm households and commute daily to their places of employment. Moreover, farmers living in areas where local non-farm employment opportunities were limited went to seek jobs in large cities far away from their farms and worked as casual unskilled labourers, for example, in construction works during the slack farming season.

As a result, the agricultural labour force declined from 11 million in 1960 to 7.3 million in 1971, a decrease of 39 per cent.[15] Although the compound annual rate of decrease was 4.4 per cent over this period, it was considerably accelerated after 1969, when the programme for curtailment of rice production became operative. Specifically, the rate of decrease was 5.9 per cent in 1970 and 9.6 per cent in 1971, as compared with 3.2 per cent annually from 1965 to 1969.

The agricultural labour force consists almost exclusively of the self-employed and family workers. There are virtually no hired regular labourers.[16] The number of new entrants into farming has been decreasing and the younger farm workers are increasingly leaving farms. Consequently the number of aged and female workers has become an increased proportion of the total number of farm workers. In 1972, for example, male workers of 60 years of age and older and females of all ages accounted for 13 and 56 per cent respectively of those who were usually engaged exclusively or mainly in agriculture. Taking both sexes together, 20 per cent were more than 60 years of age. Agriculture has, as a matter of fact, contributed to alleviation of the serious plight of the Japanese elderly for whom social welfare benefits have so far been inadequate.

*3 Material inputs and capital formation*

As in other countries, technological advance in Japanese agriculture including animal husbandry has required the increased use of purchased inputs. According to the Agricultural Sector Account compiled by the Ministry of Agriculture and Forestry, the current operating expenses of the agricultural sector at 1970 prices amounted to 1434 billion yen in 1971 or 2.7 times as much as in 1960. In particular, expenditures for feeds and agricultural chemicals more than quadrupled during this period. During most of the past decade, the price index for agricultural requisites rose rather moderately in comparison with agricultural prices (Table 1.11).

TABLE 1.11 Price indices for agricultural produce, requisites and paddy land, 1960–71

| | *Index of agricultural prices* | *Index of prices of agricultural requisites (1965=100)* | *Index of paddy land prices*[a] |
|---|---|---|---|
| 1960 | 67 | 86 | 58 |
| 1961 | 73 | 90 | 68 |
| 1962 | 80 | 91 | 74 |
| 1963 | 86 | 94 | 78 |
| 1964 | 90 | 95 | 96 |
| 1965 | 100 | 100 | 100 |
| 1966 | 107 | 104 | 110 |
| 1967 | 117 | 109 | 142 |
| 1968 | 119 | 112 | 177 |
| 1969 | 127 | 112 | 225 |
| 1970 | 131 | 116 | 295 |
| 1971 | 133 | 120 | 360 |

[a] The national average price of middle-grade paddy field transacted for agricultural purposes.

Fixed capital formation in agriculture, at 1970 prices, increased by 2.5 times between 1960 and 1971, although with the restraint of rice production it has levelled off during recent years. Of late, investments in land improvement (supported by direct government investment and subsidies) absorbed the largest share of the total fixed capital formation, followed by investment in buildings for agricultural use and the purchase of farm machinery.

Rising rural wages, as well as demonstrated technical benefits, have induced farmers to introduce various types of modern machines. In 1972, for instance, 63 per cent of farmers owned power tillers, 5 per cent tractors, 25 per cent power sprayers, 32 per cent grain driers. In addition, paddy transplanting machines and mini-combines have lately been introduced in larger numbers. Mechanization has eliminated draught animals and also reduced input of human labour. According to the Rice Production Cost Survey by the Ministry of Agriculture and Forestry, the average labour input required to produce 150 kilogrammes of brown rice decreased from 58 man-hours for the 1960 crop to 36 man-hours for the 1971 crop, with notable savings of labour in ploughing, weeding, harvesting and transplanting. Because of the small size of their farms, however, many machines owned by individual farmers are actually often under-utilized. The efficient use of larger machines like tractors and combine harvesters in particular requires larger-scale holdings, or joint or cooperative use of these machines.

*4 Farm structure*

With the continuing outflow of farm population, the number of farms[17] began to decrease around 1960 (when it numbered 6,057,000) and fell to 5,170,000 in 1972 (see Table 1.12). The compound annual rate of decrease during this period was 1.3 per cent which was much lower than the rate of decrease in the agricultural labour force. The decrease in the number of farms resulted mostly from the absence of heirs to take over the farms when the farmer died or became unable to work.

The number of part-time farms, which had already been quite large, increased remarkably and accounted for 86 per cent of the total number of farms in 1972 as compared with 66 per cent in 1960. They are usually classified into two types: Type 1, part-time farmers who earn more income from their farming operations than from other jobs, and Type 2, those who earn more income from other jobs than from their farming operations. In the above-mentioned period Type 1 decreased, while Type 2 increased remarkably to represent more than two-thirds of all the part-time farms or 59 per cent of the total number of farms in 1972. In actual fact, many of the farmers are essentially regular wage-earners who also possess some farm land.

In contrast, the number of full-time farms dropped by nearly two-thirds during the same period and occupied only 14 per cent of all the farms in 1972, as compared with 34 per cent in 1960. Since these full-time farms include those operated by retired or old-age farmers alone, and since Type 1 part-time farms include large farms with a full-time male farm worker who also works away from his farm only during the slack season or whose sons or daughters commute to factories or offices, a different base for classifying full-time farms in a more realistic way might be useful – for example, to define them as farms with at least one male worker under 60 years of age who is engaged in farming on his own farm for more than 150 days a year. The number of such farms in 1971 represented 34 per cent of all farms in the country (excluding Hokkaido, the northernmost island where the size of farm is usually larger than in the rest of Japan).

In 1972 about 80 per cent of all farms, except those in Hokkaido, marketed some part of their production, the rest being subsistence farms.

Of the former, rice farms represented 60 per cent, followed by mixed farms (11 per cent), fruit farms (10 per cent), dairy and livestock farms (8 per cent) and vegetables farms (70 per cent).[18]

Another basic characteristic of Japanese farm structure is the extremely small size of individual farm holdings. On the average, only 1.1 hectares of cultivated land is operated per farm. In Hokkaido, which has some special features and where upland farming is widely practised, the average size of farms is exceptionally large by Japanese standards – nearly 6 hectares of cultivated land. The average size of farm falls to only 0.9 hectares when Hokkaido is excluded. Furthermore, these small farms consist of a number of scattered tiny pieces of land.[19] Almost 70 per cent of farms have less than

TABLE 1.12 Number and percentage distribution of farms[a] classified as full-time and part-time 1960, 1965, 1970 and 1972

| | *1960* | *1965* | *1970* | *1972* | *1972 over 1960* (%) |
|---|---|---|---|---|---|
| NUMBER OF FARMS ('000s) | | | | | |
| Full-time[b] | 2078 | 1219 | 831 | 743 | 36 |
| Part-time[c] | 3978 | 4446 | 4510 | 4427 | 111 |
| Type 1 | 2036 | 2081 | 1802 | 1404 | 69 |
| Type 2 | 1942 | 2365 | 2709 | 3023 | 156 |
| Total | 6057 | 5665 | 5342 | 5170 | 85 |
| PERCENTAGE DISTRIBUTION | | | | | |
| Full-time | 34 | 21 | 16 | 14 | — |
| Part-time | 66 | 79 | 84 | 86 | — |
| Type 1 | 34 | 37 | 34 | 27 | — |
| Type 2 | 32 | 42 | 51 | 59 | — |
| Total | 100 | 100 | 100 | 100 | — |

[a] Farms are defined on the basis of farm households which carry on farming on land of either over 10 ares in Eastern Japan or over 5 ares in Western Japan, and which, even in case of less or no land, have annual sale of agricultural products over a certain minimum (e.g. Y 50,000 in 1970 and 1972).
One hectare equals 100 ares.

[b] Farm households of which no family members are employed in non-farm work over 30 days a year or without self-employed non-agricultural business other than their own farming earning over Y 30,000 a year.

[c] Farm households other than full-time ones. Type 1 consists of those farm households whose income from their own farming is greater than that from other sources. Type 2 consists of those farm households whose income from their own farming is less than that from other sources.

1 hectare each and only 6 per cent have more than 2 hectares each. Even in the case of full-time farms, many have an inadequate area of land to cultivate. Indeed, there is no distinct difference in farm size between the full-time and the part-time farm as is generally the case in most other developed countries. In Japan, when full-time farmers become part time they usually abandon all but rice cultivation, yet seldom feel it necessary to reduce their farm size. The increase noted earlier in the number of part-time farms, therefore, has made very little if any land available for the enlargement of full-time farms. Generally speaking, full-time farms have been more inclined to take up capital-intensive lines of production requiring relatively less land, such as pig or poultry raising, vegetable growing in vinyl houses, etc. – particularly so after the policy for curtailing rice production was adopted.

A strategic aim of the Agricultural Basic Law was to foster viable farms. In this context the viable farm is defined as one earning agricultural income at least equal on a per caput basis to the income of the average wage-earner living in towns and villages. It is estimated that the proportion of viable units rose from 9 per cent in 1960 to 13 per cent in 1967 and then fell to 5 per cent in 1971.[20] Since the minimum agricultural income required for this purpose went up by 11.9 per cent annually from 1960 to 1971, it was not easy for a farm to remain in this category. The increase in the share of viable farms during the period 1960–7 is attributable to increased output and rising prices of rice and other farm products. The subsequent decrease is largely explained by the curtailment of rice production and a slowing down in the rise in farm prices. Among the viable farms in 1971, vegetable, fruit and mixed farms were apparently the most numerous, accounting for one-fifth each, followed by rice, pig and dairy farms.

Table 1.13 shows the estimated shares in 1971 of the various categories of farms in the total cultivated area and agricultural output for Japan, (excluding Hokkaido). Type 2 part-time farms accounted for 39 per cent of all cultivated land and produced 28 per cent of total agricultural output, including 38 per cent of the rice production. Farms with at least one full-time male farm worker under 60 years of age accounted for 51 per cent of the cultivated land and contributed 65 per cent of the total output. It is estimated that viable farms accounted for 13 per cent of cultivated land and 21 per cent of total Japanese agricultural production.

There has been no appreciable progress in enlarging the size of individual farm operations, except for the capital-intensive lines of production requiring relatively little land. Despite various measures taken operations the government in the last decade, there still exist serious obstacles which have proved difficult to overcome. First, the traditional employment system based on life-long employment and progress on a seniority basis as referred to earlier persists, and old age benefits to farmers under the social welfare system have been improved only very recently.

Table 1.13 Estimated share in farm land and output of different types of farms, 1971

| | Share in the total (percentages) | | | | Average size of farms | Multiple cropping ratio |
|---|---|---|---|---|---|---|
| | *Number of farms* | *Farm land under cultivation* | *Value of agricultural output* | *Of which value of rice output* | *Ares*[c] | (%) |
| By full-time or part-time farms[a] | | | | | | |
| Full-time farms | 14 | 19 | 25 | 16 | 145 | 108 |
| Type 1 part-time farms | 27 | 42 | 47 | 45 | 155 | 105 |
| Type 2 part-time farms | 59 | 39 | 28 | 38 | 66 | 98 |
| By type of farm worker[b] | | | | | | |
| Frams with at least one male full-time farm worker under 60 years old | 34 | 51 | 65 | 46 | 152 | 109 |
| Farms with only female full-time farm worker and/or male full-time farm worker over 60 years old | 18 | 19 | 17 | 20 | 108 | 103 |
| Farms without any full-time farm worker | 48 | 29 | 18 | 34 | 61 | 96 |

[a] For definition of full-time and part-time farms, see footnote to Table 1.10.
[b] Full-time farm worker means a farmer or his family member who is engaged in farming on his own farm for 150 days and over per year.
[c] One hectare equals 100 ares.

Thus, middle-age, not to speak of old-age, farmers find it very difficult to get regular and dependable off-farm jobs with satisfactory pay. Even if they should get good jobs, they will generally be unwilling to part with their farms, which represent security for old age. Secondly, transfer of farm land is relatively limited because of the high and soaring land prices, as noted earlier, which greatly strengthens the propensity of farmers to keep their farm land as an asset of increasing value. Moreover, transactions in pieces of farm land have been carried out mostly between farmers of the same or similar groups, and relatively few have contributed to enlargement of farm size. Thirdly, few landowners lease their land to other farmers because of the strong security of tenure rights established under the post-war land reform. The leasing of farm land has tended to decrease, and represented only 5 per cent of the total cultivated land in 1970. It is reported however that there has been some increase in arrangements whereby a farmer undertakes full operation of other farmers' land on a commission basis.

In view of the many difficulties involved in enlarging the size of individual holdings, farmers have moved to consolidate some of their farming operations with other farmers. The number of joint farm enterprises is however small and gradually diminishing. In rice farming, there are three broad categories of group farming: (a) joint farming operations; (b) joint use of large machinery, equipment and other facilities; and (c) arrangements whereby some farming operations are entrusted to a small group of full-time farmers or local agricultural cooperatives. The adjustment of rice production in recent years has adversely affected the development of group farming units. It appears, however, that those which have been farmed have tended to evolve from category (a) to (b) and then to (c). In other fields, such as fruit growing, vegetable growing, sericulture, there are similar group farming units formed primarily for joint use of machinery and of equipment (for example, for controlling pests and diseases and nursing baby silkworms) as well as for the marketing of produce. It is estimated that in 1972 group farming units covered 26 per cent of the total area planted to rice, 22 per cent of the total area of orchards, and 37 per cent of the total silkworm cocoons gathered.

### *5 Farm income levels*

Among the different types of farming, rice farmers had generally fared better until around 1968, due to the increased yield per hectare and the higher producer prices. Since then, however, their situation has tended to worsen as a result of acreage reduction and the pegging of prices. As can be seen in Table 1.14, which shows the average agricultural income per farm by types, the level of agricultural income in 1971 of the average vegetable, fruit, dairy, pig or poultry farm wage-earner's that of the average rice

TABLE 1.14 Average agricultural income per farm by type of farming,[a] 1971

| | *Rice farm* | *Vegetable farm* | *Fruit farm* | *Sericulture farm* | *Poultry farm* | *Pig farm* | *Dairy farm* |
|---|---|---|---|---|---|---|---|
| Agricultural income (Y thousand) | 436 | 989 | 835 | 327 | 734 | 902 | 1026 |
| Its relative level (Rice farm = 100): | | | | | | | |
| *1967* | 100 | 109 | 92 | 63 | 97 | 148 | 121 |
| *1971* | 100 | 227 | 192 | 75 | 168 | 207 | 235 |
| Agricultural income as percentage of farm income: | | | | | | | |
| *1967* | 60 | 73 | 64 | 54 | 63 | 56 | 80 |
| *1971* | 28 | 60 | 48 | 23 | 44 | 52 | 67 |
| Percentage of family members' labour hours spent on their own farming | 40 | 77 | 57 | 48 | 62 | 58 | 76 |
| Number of full-time farm worker | 0.5 | 2.0 | 1.0 | 0.9 | 1.1 | 1.3 | 2.0 |
| Net output per 10 hours of farm work (Y) | 2198 | 1967 | 2929 | 1231 | 2119 | 2693 | 2035 |
| Size of planting or herd of principal line of production | *hectares* 1.0 | *hectares* 0.7 | *hectares* 0.3 | *Cases*[b] 13 | *Birds*[c] 1470 | *Head*[d] 72 | *Head*[e] 8 |
| Percentage of sales proceeds from principal product | 84 | 75 | 75 | 62 | 87 | 68 | 84 |

[a] Farms are reclassified according to their principal line of production which earns more than 60 per cent of the total cash receipts from farming.
[b] Annual cases for gathering silk worms from egg paper.
[c] Monthly average number of laying hens kept.
[d] Annual number of pigs sold.
[e] Monthly average number of cows in milk.

TABLE 1.15 Income comparison between farm and wage-earners' families, 1971
(in thousands of yen)

| | *Number of family members* | *Total income* | *Of which agricultural income* | *Per capita income* | *Income ratio (wage-earner) (=100) (%)* |
|---|---|---|---|---|---|
| Average wage-earners' family | 3.88 | 1521 | — | 392 | 100 |
| | (4.17) | (655) | | (157) | |
| Average farm family | 4.74 | 1780 | 501 | 376 | 96 |
| | (5.43) | (642) | (289) | (118) | (75) |
| By type of farm worker | | | | | |
| Farm family with at least one male full-time farm worker under 60 years old | 5.25 | 1721 | 1003 | 328 | 84 |
| Farm family with only female full-time worker and/or male full-time farm worker over 60 years old | 4.83 | 1697 | 503 | 351 | 90 |
| Farm family without any full-time farm worker | 4.48 | 1830 | 202 | 409 | 104 |

*Note*: Figures in parentheses relate to the year 1963.

farm. The sericultural farm remained with the lowest average agricultural income.

Virtually all Japanese farms are family enterprises dependent on family labour and with farm operations not distinctly separable from those of the household. Farmers' incomes from all sources showed a marked increase during the 1960s. According to the Farm Household Economy Survey by the Ministry of Agriculture and Forestry, the national average farm household income at current prices in 1971 totalled 1,780,000 yen, or almost three times as much as in 1963 (642,000 yen). Of the total, only 28 per cent came from its own farming operations and the remaining 72 per cent from other sources. The latter ratio compares with 55 per cent in 1963. In other words, the increase in farm income was attributable mainly to the expansion of non-agricultural income, which, in turn, reflected both an increase in the number of part-time farms and rising wage levels in general.

Table 1.15 gives a rough comparison of the incomes of farm and wage-earning families.[21] The per caput income for the average farm household amounted to 376,000 yen in 1971, equal to 96 per cent of that for the national average wage-earner's household. The corresponding percentage in 1963 had been 75. The income level for part-time farms improved to a greater extent than did that for the average farm household. The per caput income of farm households with no member engaged in farming for over 150 days a year averaged 409,000 yen in 1971, 4 per cent above that of the average wage-earner's family. By contrast, farms with at least one male full-time farm worker under 60 years of age earned on the average a total income of 1,721,000 yen in 1971, of which around 40 per cent came from non-agricultural sources. On a per caput basis it was equivalent to 328,000 yen or 84 per cent of the per caput income for the average wage-earner's family.

Thus the primary aim of income parity set forth in the Agricultural Basic Law has on the whole been attained. However, the main contributing factor has not been greater income from farm operations, as anticipated by that law, but rather, much larger non-agricultural income from off-farm sources. The more the farm family has depended on farm operations for its income, the worse off it has been in relative terms, and among various categories of farmers genuine full-time farmers have the lowest average income per caput.

## NOTES

1 Including both full- and part-time labourers, with due allowances for part-time labourers.
2 Ministry of Agriculture and Forestry, 'Analysis of Food Demand, 1971', p. 7 (in Japanese).
3 The combined share of wheat, barleys, miscellaneous grains, pulses and

potatoes in the value of total agricultural output fell from 12 per cent in 1960 to 4 per cent in 1971.

4 The crop insurance scheme covers rice, wheat, barleys, cocoons and livestock.

5 Some of the average rates of subsidy per one-tenth of a hectare were as follows:

| | |
|---|---|
| Suspension of rice cultivation | Y 30,000 |
| Ordinary diversion (i.e. individual diversion to other annual crops) | Y 35,000 |
| Special diversion (i.e. individual diversion to perennial crops or group diversion to other crops) | Y 40,000 |

For reference, it is estimated that the average net income from rice growing for the 1971 crop amounted to approximately Y 40,000 per one-tenth of a hectare.

6 In addition, the combined imports of forest and fishery products amounted to as much as US $2060m in 1971 compared with US $204m in 1960.

7 Computed from FAO, *Trade Yearbook*, 1971.

8 The import of maize for other uses also increased sharply, amounting to 1.6 million tons in 1971, which substituted in part for domestic sweet potatoes as a raw material for producing starch.

9 Imports of cotton yarn and fabric from Asia and the Far East were negligible in 1960 but totalled nearly US $40m in 1971. Similarly, imports of jute yarn and fabric from the same region amounted to approximately US $6m in 1971 as against US $0.4m in 1960.

10 The number of items is based on the SITC four-digit classification.

11 However, it should be noted that this protection has not necessarily brought about large domestic production and small imports. For instance, both wheat and barley registered a sharp decline in production and a marked expansion of imports, as stated earlier.

12 The self-sufficiency ratio for all foods (including those of fishery origin) is estimated at 77 per cent for 1971 as against 93 per cent for 1960.

13 The self-sufficiency ratio based on 'original calories' means that all foods are measured in terms of calories, with livestock products being measured in terms of calories of feeds estimated to have been used for their production. Thus livestock production, to the extent that it depends on imported feeds, is virtually treated not as domestic production but as imported.

14 The stimulus to imports of agricultural products arising from the revaluation of the yen has been offset to a considerable extent since it has coincided with soaring commodity prices in the world market.

15 Prime Minister's Officer, *Labour Force Survey* (various dates). In this monthly survey, the actual employment status approach is used. The agricultural labour force includes 'the employed at work' and 'the employed not at work'. The former means all persons who worked for pay or profit for at least one hour during the survey week. Unpaid family workers are also included in this category. The latter denotes self-employed workers who did not work during the survey week but whose absence from work has not exceeded 30 days. If a person has two or more jobs, he is classified according to his primary job, i.e. a job on which he worked the longest time during the survey week.

16 In 1970 fewer than 18,000 farms employed one or more regular labourers, of whom presumably some were actually in-service trainees.

17 For the definition of farms, see footnote of Table 1.12.

18 Ministry of Agriculture and Forestry, *Annual Sample Census of Agriculture, 1971*. Farms are classified according to their principal line of production which earns more than 60 per cent of the total cash receipts from farming.

19 Except in Hokkaido, the average farm was composed of five to six scattered small tracts of land averaging 14 acres each (*Source*: a survey conducted by the Ministry of Agriculture and Forestry in 1953).

20 Government of Japan, *Annual Agricultural Report*, 1972.

21 For stricter comparison, account should be taken of price differentials, different treatment of dwelling rent, etc., which are ignored here. In the Farm Household Economy Survey, farm products produced and consumed by the same farm household are valued at prices received by farmers, which are mostly lower than retail prices. In regard to rice which accounts for some three-fourths of the value of such home-consumed farm products, however, producer prices have been higher than retail prices as a result of government control as stated earlier. In the same survey farm building is treated as an asset and therefore depreciation cost is counted. As to retail prices of consumer goods, it is difficult to determine the precise differential between rural and urban areas because the type and description of commodities adopted in the price surveys often differ between rural and urban areas. It appears that in rural areas some consumer goods are cheaper while others, particularly durable ones, are more expensive, and that the over-all price differential, if any, is perhaps minor.

# 2 The Prospects up to 1982

## OBJECTIVES AND CHARACTER OF MAF PROJECTIONS

In October 1972 the Japanese Ministry of Agriculture and Forestry (MAF) adopted 'Projections of Demand for and the Production of Agricultural Products for 1982' to provide guidelines for the government's agricultural policy during the coming decade. The ultimate purpose of the MAF projections was to ensure a sound development of Japanese agriculture, through improving productivity and the use of land resources. The aims were efficient as well as secure sources of supply of good-quality food to meet Japan's domestic demand. The demand for major agricultural products in 1982 was projected to this end, and indicators of desired improvements in labour productivity set forth. Production targets were established as the long-range goals for 1982, so that the implications of future production trends could be adequately assessed and necessary policy changes introduced.

The projections are not simply extrapolations of past trends, but take account also of the intentions of the government with respect to future agricultural production. In other words, the Ministry of Agriculture and Forestry has set production targets with indicators of desired improvements in productivity and target self-sufficiency ratios for major agricultural products. The Ministry considered it important to reveal its intention regarding the future orientation of agricultural production to provide a guideline for farmers and others.

For the purpose of projection, the year 1970 was adopted as the base and 1982 as the target year. Practically all the major agricultural products were covered: rice, wheat, and barley, sweet and white potatoes, pulses, green tea, sugar, vegetables, fruits, raw silk, dairy products, meat, eggs and feedstuffs, which altogether represent approximately 95 per cent of the total domestic agricultural output.

The demand projections were based primarily on correlation analysis between per caput personal consumption expenditures and the quantity of per caput consumption in the past. Selected types of demand functions were used to forecast the target year consumption, and informed judgement was applied with respect to some points. Compound annual rates of growth of personal consumption expenditure in real terms during 1970 to 1982 were assumed as 7, 8 or 9 per cent,[1] each representing a hypothesis of low, medium or high rate of growth.

The future population growth rate was based upon the result of a study made by the Institute of Population Problems of the Ministry of Welfare. Accordingly, the population was projected to increase from 103.7 million in 1970 to 118.0 million in 1982 at the rate of 1.1 per cent per annum.

It was assumed that relative prices would continue generally in line with past trends. As a working hypothesis, the base period situation with respect to international trade arrangements was taken as a first approximation. As for future developments in international commodity markets, it was, broadly speaking, presumed that the general picture given in FAO's 'Agricultural Commodity Projections, 1970–1980' (Rome, 1971) would materialize.

For establishment of the production targets, simple demand and production projections were first matched, and the results then adjusted to take account of a comprehensive study of commodity features, trends to demand, possibilities for increasing both domestic production and productivity, stability of overseas supply, competitive position of domestic production *vis-à-vis* imports, etc. Considerable attention was also given to the influence of implied production adjustment on farm management, the position of agriculture in the various regions of the country, effective utilization of resources, security of agricultural incomes, conservation of natural environment, etc. The results of the MAF projections are outlined in the following sections.

## DEMAND PROJECTIONS FOR AGRICULTURAL PRODUCTS

As mentioned above, food demand for 1982 was projected on the basis of three possible rates (7, 8 and 9 per cent) of increase in real personal consumption expenditure. To take the 8 per cent rate as an example, food expenditure will most likely increase by 5.3 per cent annually, with tendencies towards more protective foods and greater variety. By 1982, for instance, the per caput consumption of meat would be doubled, that of dairy products would show a 50 per cent increase, and that of fruits and vegetables would expand by approximately 40 and 20 per cent respectively. The per caput consumption of fats and oils and sugar would also increase by about 40 and 20 per cent respectively, while that of eggs would show only a modest growth since the level of present consumption is already quite high. During the same period, on the other hand, rice consumption per caput would continue to decrease by about 20 per cent, to 75 kilogrammes (milled basis) per year in 1982, and even wheat would show a slight decline.

As a result, the national average intake of calories per caput per day in 1982 would be about 2640 (a 7 per cent increase over 1970) and that of protein and fat 90 grammes and 74 grammes respectively (17 and 42 per cent greater respectively than in 1970). The ratio of starchy foods to the total calorie intake will continue to fall (to 43 per cent in 1982), while that of animal protein to the total protein intake will substantially rise to 53 per cent. It is expected that the unique dietary pattern will continue, however, and that relatively more starchy foods, as compared with the pattern of most Western countries, will continue to be consumed.

The total food demand in 1982, in terms of real food expenditure on the basis of the medium rate of growth assumption (8 per cent), would indicate an increase of almost 90 per cent from the 1970 level. The percentage increase in demand between these two years would be 131 for meat (with total demand amounting to 3.9 million tons), 72 for dairy products (9.2m tons in fresh milk equivalent), 62 for fruits (10.7m tons), 40 for sugar (4m tons), 39 for vegetables (21m tons), 35 for soybeans (4.4m tons), 25 for eggs (2.3m tons), and 11 for wheat (5.8m tons). On the other hand, the total demand for rice would decline by nearly 10 per cent (to 10.8m tons in terms of brown rice in 1982) and that for sweet potatoes would be almost halved during the same period. The total demand for raw silk would increase by 59 per cent (to 648,000 bales in 1982).

## PROJECTIONS FOR AGRICULTURAL PRODUCTION

### *1 Likely farm structures*

In projecting agricultural production, the following assumptions were made as to structural factors affecting production. First, the area of cultivated land would decrease from 5.8 million hectares in 1970 to 5.2 million hectares in 1982, the area of existing cultivated land diverted to industrial, residential and road sites, etc., would exceed the area of land newly brought under cultivation of for example grass, vegetables, fruits, etc. In response to expanding food requirements, on the other hand, permanent grassland would increase sharply from 237,000 hectares in 1970 to 637,000 hectares in 1982.

The downward trend in the agricultural labour force would be accelerated, as a result of the decentralization of factories into rural areas and of increased retirement of elderly farmers. Thus the agricultural labour force would decline by approximately 5.0 per cent annually to some 4.3 million, or roughly half of the 1970 level. Of these, the elderly farmer group would increase relatively more.

The number of farms would also decrease but at a much slower rate. It would be likely to decline by 1.7 per cent per annum to a little more than 4.3 million in 1982, or approximately 80 per cent of the number in 1970. It is anticipated that Type 2 part-time farms would represent about 60 per cent of all farms in 1982. With continuing labour shortage and rising wage levels, there will be strong preference for completely mechanized forming systems using big or medium-sized farm machinery.

In order to attain the production targets it is expected that various

TABLE 2.1 Demand projections for major agricultural commodities for 1982[a]

| Commodity | 1970 (Actual) | | 1982 (projection)[b] | |
|---|---|---|---|---|
| | Total demand (1000 tons) | Per caput net food supply (kg) | Total demand (1000 tons) | Per caput net food supply (kg) |
| Rice | 11,948 | 95·1 | 11,090–10,830–10,556 | 77·3–75·3–73·3 |
| Wheat | 5207 | 30.8 | 5760 | 29·4 |
| Barley | 1685 | 1·5 | 2310–2340–2360 | 0.9 |
| Sweet potatoes | 2564 | 4·1 | 1473 | 4·5 |
| White potatoes | 3611 | 12.1 | 3607 | 11.9 |
| Soybeans | 3281 | 5.6 | 4427 | 5.8 |
| Peanuts | 115 | 1.0 | 118 | 1·5 |
| Other pulses | 397 | 3·3 | 381 | 2.9 |
| Tea | 100 | 0.95 | 127–130–132 | 1.06–1.08–1.10 |
| Vegetables | 15,210 | 115·5 | 20,840–21,170–21,474 | 138.6–140.7–142.6 |
| Fruits | 6636 | 38.2 | 10,273–10,736–11,195 | 51·9–54·3–56·7 |
| Milk and milk products | 5355 | 50.1 | 8751–9230–9705 | 72.2–76.2–80.2 |
| Meat[c] | 1678 | 11.6 | 3555–3878–4220 | 21.4–23·3–25·5 |
| Eggs | 1817 | 14.8 | 2271 | 16.3 |
| Sugar | 2829 | 26.8 | 3820–3974–4117 | 31.8–33.1–34·3 |
| Fats and oils | 1361 | 9·5 | 2030–2088–2141 | 13.0–13·4–13·9 |
| Raw silk[d] | 408 | | 611–648–686 | |

[a] Fiscal year.

[b] Based on assumed annual average in three variants according to the assumption of annual average growth rates of personal consumption expenditure in real terms at 7, 8 and 9 per cent respectively.

[c] Excluding whale meat.

[d] In terms of 1000 bales of 60 kg each.

policy measures already in effect under the Agricultural Basic Law would be continued and, wherever necessary, further strengthened. In this connection, special emphasis was placed on the need to enlarge the scale of farming operation through group farming, principally by full-time farmers utilizing resources belonging to part-time farms. Broadly speaking, it is expected that there would be further concentration of pig or poultry raising, which requires little land, with large-scale specialized producers further enlarging their herds or flocks and accounting for a dominant share in commercial output. In dairy farming, specialized producers with 40 to 50 cows each are expected to increase in number and account for one-half of the total population of dairy cattle. As for the breeding of beef calves, farms with herds of 10 to 15 head of beef cattle, carrying on mixed farming with rice or other crops, are expected to become prevalent, though in the case of cattle fattening a considerable number of feedlots fattening dairy steers are also expected to come into existence. As regards rice growing, on the other hand, various cultural operations on part-time farms will increasingly be undertaken by individual full-time farmers or by group farming units composed of full-time farmers; it is anticipated that the latter will cover as much as 60 to 70 per cent of the total area planted to rice. Likewise, group farming units for pest control purposes are expected to cover roughly half of the total output of fruits; and the expansion of group farming is also expected for sericulture, particularly for the rearing of baby silkworms and the cultivation of mulberry fields.

*2 Production and productivity*

The index of agricultural production in 1982 will according to the projections show a 31 per cent increase over 1970, which is equivalent to a compound rate of growth of 2.3 per cent per annum. During this period livestock production will increase by as much as 91 per cent. The largest increases are expected in pigmeat and broiler production, which will be more than doubled. The increase in beef and dairy production will be rather less. The production of fruits and vegetables will increase by 62 and 42 per cent respectively. Sericulture production will also increase greatly; and even the output of soybeans, peanuts and malting barley will grow substantially. By contrast, the production of rice and sweet potatoes will decrease by 15 and 43 per cent respectively.

In consequence, there will be a remarkable change in the commodity composition of the total agricultural output. In 1982 the share of livestock products, vegetables and fruits will go up to 32, 16 and 10 per cent respectively, while that of rice will fall to 24 per cent.[2] Among livestock products pigmeat will represent by far the biggest share, followed by milk and eggs.

In 1982 the area planted to rice will decrease to 2.2 million hectares, or nearly 0.7 million hectares less than in 1970. The area under wheat, non-malting barley and potatoes will also diminish significantly, while the area of vegetables and fruits will increase substantially. In addition, the area under soybeans will recover to 240,000 hectares, or two and a half times as much as in 1970. Also, the area under forage, mainly grass, will expand by

TABLE 2.2 Production projections for agricultural commodities for 1982[a]

| *Commodity* | *Production* 1970 (Actual) | *Production* 1982 (Projection) | *Index 1982* |
|---|---|---|---|
| | (1000 tons) | | (1970 = 100) |
| Rice | 12,689 | 10,830 | 85.3 |
| Wheat | 474 | 480 | 101.3 |
| Barley | 572 | 578 | 101.0 |
| Sweet potatoes | 2564 | 1473 | 57.4 |
| White potatoes | 3611 | 3607 | 99.9 |
| Soybeans | 126 | 536 | 425.4 |
| Peanuts | 70 | 94 | 134.3 |
| Other pulses | 255 | 222 | 87.1 |
| Tea | 91 | 130 | 142.9 |
| Vegetables | 15,126 | 21,170 | 140.0 |
| Fruits | 5454 | 8827 | 161.8 |
| Milk and milk products | 4789 | 8482 | 177.1 |
| Meat[b] | 1473 | 3445 | 234.6 |
| Beef | 265 | 500 | 188.7 |
| Pig meat | 691 | 1798 | 260.2 |
| Chicken | 505 | 1157 | 229.1 |
| Eggs | 1766 | 2271 | 128.5 |
| Sugar | 642 | 1058 | 164.8 |
| Raw silk[c] | 342 | 529 | 154.7 |

[a] Fiscal year.
[b] Excluding whale meat.
[c] In terms of 1000bales of 60 kg each.

one-half to 992,000 hectares in 1982, which coupled with 637,000 hectares of permanent grassland will supply about 8.9 million tons (calculated in terms of TDN)[3] of herbage, or some 60 per cent of the total feed requirements for dairy and beef cattle. On balance, the total planted area in 1982 will decline to 5.9 million hectares on 0.4 million hectares smaller than in 1970, although the multiple cropping index will go up again to 114 per cent.[4]

Improvements in productivity to be sought during the period are shown in Table 2.3. Chiefly as a result of expanding the size of herds and flocks, output per man-hour is to be more than trebled in the production of fat cattle, broilers and eggs, and more than doubled in that of pigs. Improved forage production and enlargement of herds is also to double productivity

TABLE 2.3 Prospective agricultural labour productivity

| *Commodity* | *Productivity per unit* *Working hour* | *Output* | *Output per working hour* | *1982/1970* | *Means for improvement in productivity* |
|---|---|---|---|---|---|
| | *Hour/10are*[a] | *KG/10are*[a] | *Kg/hour* | *Times* | |
| Rice | 65–75 | 490 | 6.5–7.5 | 1.8–2.0 | Promotion of group farming units, utilizing high-efficiency machines. |
| Wheat | 25–35 | 318 | 9.1–12.7 | 2.5–3.5 | Organization of group farming mainly by making use of land left idle after rice cropping. |
| Barley | 20–30 | 356 | 11.9–17.7 | 2.2–3.3 | |
| Soybeans | 27.4 | 223 | 8.14 | 2.3 | Establishment of rotation system on upland fields and organization of group farming. |
| Mandarin oranges | 140 | 3,640 | 26.0 | 2.2 | Organization of group farming with emphasis on pest control. |
| Apples | 153 | 2,540 | 23.1 | 2.5 | |
| Cocoons | 154 | 87 | 0.57 | 2.7 | Organization of group farming with emphasis on rearing of young silkworms and taking care of mulberry fields. |
| | *Hour/head* | *Kg/head* | *Kg/head* | | |
| Dairy farming | 201 | 5,476 | 27.3 | 2.2 | Improvement of production infrastructure and expansion of scale of breeding. |
| | | *(Head)* | *(Head/100 hrs)* | | |
| Beef cattle (for breeding) | 161 | 1.0 | 0.7 | 2.3 | |
| Beef cattle (for fattening) | 113 | 310 | 2.7 | 3.4 | |
| Pigs for fattening | 6.8 | 65 | 9.6 | 2.3 | Expansion of scale of herd of flock kept. |
| | *Hour/100 hens* | *Kg/100 hens* | *Kg/hour* | | |
| Broilers | 3.7 | 14.3 | 38.6 | 3.9 | |
| Eggs | 51 | 1,534.0 | 30.1 | 3.7 | |

[a] A measure equal to 100 sq. m.

in dairy farming and calf breeding. Output per man-hour will almost double for rice growing, largely through widespread group farming, and more than double for fruits and sericulture, partly through cooperative or group activities in some of the production processes. For upland crops like potatoes and pulses, output per man-hour will likewise rise, largely through improved practices and mechanization. Labour productivity of agriculture as a whole will in 1982 be approximately two and a half times the 1970 level.

### TARGETS FOR SELF-SUFFICIENCY RATIOS AND IMPLIED IMPORT REQUIREMENTS

The MAF projections have set production targets for major agricultural commodities in terms of both quantity and of self-sufficiency ratios. Commodities can be classified generally into four categories, in terms of the desired degree of self-sufficiency.

First, rice is the only commodity for which complete self-sufficiency is envisaged. This will, however, involve a 15 per cent reduction in rice production compared with the base year.

The second category includes commodities for which it is aimed to be largely self-sufficient. These include vegetables, fruits, dairy products, meat, eggs and raw silk. Of these, almost complete self-sufficiency is to be sought in vegetables and eggs; many vegetables are difficult to transport over long distances from overseas due to their perishability, and for eggs a sufficient improvement in productivity can reasonably be anticipated through enlargement of the scale of production enterprises. It is anticipated that production of pigmeat and broilers can also be raised relatively easily by enlarging the size of herds and flocks, and the targets have been set at 1.8 million tons and 1.2 million tons respectively. The target for beef production has been set at 0.5 million tons; in view of likely shortage on the international market, every possible effort is to be made to increase domestic beef production. Thus, the production target for all meat totals about 3.5 million tons, or roughly 90 per cent self-sufficiency. Although complete self-sufficiency is aimed at in the case of milk for liquid consumption (which requres a high degree of freshness), the intended self-sufficiency ratio for all dairy products does not differ much from that of the base year (89 per cent). Likewise, for fruits as a whole it is aimed to maintain much the same self-sufficiency ratio as in the base year (84 per cent); many indigenous fruits can hardly be supplied from foreign sources due to differences in species or varieties. Furthermore, mandarin oranges are considered to have strategic importance in furthering selective expansion of agricultural production. For raw silk also it is intended to maintain much the same self-sufficiency ratio as in the base year (85 per cent); sericulture is of vital importance in most of Japan's hill districts.

The third category includes commodities for which self-sufficiency ratios of about 50 per cent are intended, and includes pulses other than soybeans. Although the self-sufficiency ratios for these will decline as compared with the base year, they are important crops for rotation purposes in upland farming areas.

TABLE 2.4 Self-sufficiency ratio and import requirements of agricultural commodities for 1982

| *Commodity* | *Ratio of self-sufficiency* 1970 | 1982 (*percentage*) | *Net imports* 1970[a] | *Net import requirements* 1982[a] ('000 *tons*) |
|---|---|---|---|---|
| 1 Completely self-sufficient | | | | |
| Rice | 106 | 100 | −760 | 0 |
| 2 Mostly self-sufficient | | | | |
| Vegetables | 99 | approx. 100 | 86 | 0 |
| Eggs | 97 | approx. 100 | 51 | 0 |
| Meat[b] | 88 | 82–97 | 204 | 100–765 |
| Fruits | 84 | 81–88 | 1046 | 1446–2368 |
| Milk and milk products | 89 | 87–97 | 561 | 269–1223 |
| Raw silk | 87 | 80–90 | 82[c] | 82–157[c] |
| 3 Approximately 50 per cent self-sufficient | | | | |
| Pulses (other than soybeans) | 63 | approx. 55 | 233 | 253 |
| 4 Mainly dependent on imports but partially self-sufficient | | | | |
| Wheat and barley | 15 | approx. 15 | 5646 | 7042 |
| Soybeans | 4 | approx. 10 | 3244 | 3891 |
| Sugar | 23 | 26–28 | 2187 | 2762–3059 |
| Concentrated feedstuff | 33 | 20 | 9266[d] | 20,265[d] |

[a] Calculated for fiscal year on a Food Balance Sheet basis.
[b] Excluding whale meat.
[c] In terms of 1000 bales of 60 k each.
[d] In TDN (total digestible nutrients) equivalent.

The fourth category includes wheat, barley, soybeans and sugar, the requirements for which during the base period were met mainly by imports. The self-sufficiency ratios desired in 1982 include: wheat 8 per cent, barley 25 per cent, soybeans 10 per cent and sugar 27 per cent. In comparison with the present ratios, it is much higher for soybeans and somewhat higher for sugar. Indigenous soybeans are not only suitable for traditional food uses but are also officially recommended as a substitute for rice in the programme for diverting riceland to other uses. Sugarbeet and cane are of special significance in remote producing areas.

Thus, except for soybeans and sugar, the intended self-sufficiency ratios either remain much the same as in the base year or decline (Table 2.4). The over-all self-sufficiency ratio for the commodities projected is estimated at approximately 75 per cent. If calculated in terms of 'original calories',[5] however, it would be about 46 per cent.

Table 2.4 indicates the balance between the projected demand and production in 1982, indicating also likely import requirements. For practically all of the commodities listed, import requirements (in terms of quantity) projected for 1982 are larger than actual imports during the base year. For instance, imports of wheat, barley, soybeans and sugar will amount to 5.3, 1.8, 3.9 and 2.9 million tons respectively. Following the conspicuous expansion of livestock and poultry production, moreover, requirements of concentrated feedstuff in 1982 will total 25.5m tons (in TDN equivalent) as compared with 13.7m tons in 1970. Of this amount, 80 per cent or 20.3m tons will have to be imported from abroad; by comparison, the corresponding figure in 1970 was 9.3m tons. If converted into actual weight of feed grains, the 20.3m tons in TDN equivalent will be equal to nearly 30m tons. Thus, it is roughly estimated that imports of food and feedstuff required in 1982 will be almost double the 1970 level. In other words, the compound rate of their increase between 1970 and 1982 will be 6 to 7 per cent per annum.

NOTES

1 Under the new Basic Plan for Economic and Social Development (1973–7) adopted by the Government of Japan in February 1973, the annual rate of growth of personal consumption expenditure in real terms is 9.1 per cent, which if recalculated over a period of 1970 to 1977, will be equivalent to 8.7 per cent. As far as the demand and supply of agricultural products are concerned, the plan has been based on the results of MAF projections.
2 For the commodity composition in 1971, see page
3 Total digestible nutrients.
4 See Table 1.5.
5 See note 13 on p.

# 3 Summary and Conclusions

## PAST ADJUSTMENT EXPERIENCE

The setting for Japan's agricultural adjustment experience since the end of the Second World War has been largely dominated by the very rapid rate of growth which has characterized the national economy. As personal incomes have risen, the demand for food has increased strongly; total expenditures for food, at constant prices, virtually doubled between 1960 and 1971. There were, moreover, distinct shifts in the pattern of Japanese food consumption, especially from rice to livestock products and fats and oils. Domestic production was adjusted significantly to meet these changes in demand. Rice production was reduced in absolute terms and the share in total farm output fell from 48 per cent in 1960 to 35 per cent in 1971, while livestock production was almost tripled and rose to account for 26 per cent of total domestic farm production.

With a very limited area of land suitable for agricultural use and with increased competition from other sectors for available labour, Japan became increasingly dependent on imports of food and agricultural commodities to fulfil its domestic consumption requirements. Ample supplies available on world markets during this period and Japan's strong balance of payments position tended also to favour increased dependence on imports. Japan's self-sufficiency in food of agricultural origin (measured in value terms at constant prices) fell from 90 per cent in 1960 to 72 per cent in 1971; if measured in terms of 'original calories' the fall in this self-sufficiency ratio was even greater, from 78 to 53 per cent.

As can be expected under such circumstances, the relative position of the agricultural sector in the Japanese economy has been dramatically reduced. The agricultural labour force declined from 27 per cent of the total labour force in 1960 to only 14 per cent in 1971. Although the sector's contribution to net national product at current prices rose, its contribution to the national total dropped from 10 per cent in 1960 to less than 5 per cent in 1971. Thus, despite the relatively high level of capital investment in equipment and purchased inputs which together with improved technology increased yields per hectare and per head of livestock, it appears that the productivity of labour engaged in farming remains only about one-third of that in the rest of the economy. This low productivity reflects the very small size of Japanese farm holdings (on average only 0.9 hectares, excluding Hokkaido where farms are notably larger than the national average) and the fragmentation of the holdings.

The outstanding feature of the change in the structure of Japanese agriculture that has taken place during the period under review has been the greatly increased importance of part-time farming. Under a strict definition of full-time farms, their number dropped from 2.1 million in 1960 to 0.7 million in 1972, while the number of part-time farms rose from 4.0 million to 4.4 million (from 66 to 86 per cent of the total farms). In the latter year, it is estimated that part-time farms provided about three-quarters of the total Japanese agricultural output. Although under more liberal definition of full-time farms this share might be only one-half, an extraordinarily large part of Japanese agriculture has nevertheless come to consist of part-time farming. The off-farm employment of part-time farmers and other members of farm households has contributed the bulk of the increase in total incomes of farm households during the last decade.

With the sustained rapid rate of economic growth in Japan, a major agricultural policy issue has been how to raise farm incomes in step with the rapid increase in those of families in other sectors. The answer has been, in part, to facilitate the transfer of people out of agriculture; in part, to encourage enlargement of the scale of farming operations. The over-all result has been the achievement of parity of farm and non-farm incomes, and by 1971 the total income of the average Japanese farm household was even somewhat above that of the average non-farm, wage-earning family.

As to the three ways noted above of raising farm incomes, the transfer of labour to employment in other sectors has been sustained at a relatively rapid rate and has been accelerated in recent years. The policy of raising producer prices has, however, encountered problems, especially with respect to rice. Since 80 to 85 per cent of all Japanese farms produced rice and Japan had been a major rice importer, higher producer prices for this crop offered a relatively simply and direct means of generating and distributing additional income to farm households. The higher prices however brought a sharp increase in rice production and the accumulation of relatively large surplus stocks in the hands of the government. The government reacted by curbing rice production by means of direct and generous payments to farmers for taking land out of paddy production, and a prompt downward adjustment in rice production followed.

Increases in the scale of farming operations have, however, been limited by the lack of land available for enlargement of farm holdings. Between 1960 and 1971 the average price of middle-grade paddy land rose by more than six times, and even part-time farmers who could not make full use of their land have preferred to retain it for its increasing value as an asset. Owners of land have also been reluctant to rent it because of the security of tenure which accrued to tenants under the post-war land reform. Thus, increases in the scale of farming operations have been limited largely to farms producing commodities (for example, pigs, broilers, eggs) for which the production could be expanded by increasing capital investment in livestock, equipment and purchased feeds without increasing the land area of the farm holding.

## FUTURE PROSPECTS

### *1 Structural aspects*

Although some measure of structural adjustment has already been achieved, Japan faces the need to undertake further changes, possibly of a somewhat different nature. The basic policy issue appears to be what sort of farming and what agricultural policies are best suited to generally small-scale and, to an increasing degree, part-time farming in a thickly populated, high-income country. The appropriate mix of small part-time farms whose operators earn most of their income elsewhere and of largely full-time farms is likely to be influenced by the priority accorded to maximizing the productivity of land as well as by the role given to imports as a source of food and feed.

With the expansion of part-time farming especially, Japanese agriculture has become increasingly integrated into the national economy and no longer constitutes a sharply defined sector, either economically or socially. Furthermore, agriculture presently provides less than one-third of the total incomes of Japanese farm families, the number of which is expected to continue to decline. Thus in the future evolution of Japanese agriculture policies it will be possible for relatively greater attention to be given to the needs of the economy as a whole and relatively less to the need to improve incomes for a special segment of the population.

If Japanese farming is to be economically viable with domestic producer prices at levels comparable to the cost of imported supplies, there must be an enlargement of the size of farms. This implies either policies of land redistribution (which might not prove politically acceptable), a substantial spread of group farming (which, although the official objective, has not actually progressed very far), or a variety of coordinated and strong policy measures such as a strict regulation of land use, greater incentive for *de facto* leasing of farm land, urbanization and industrialization carefully planned with total land use considerations in mind, and so forth.

### *2 Commodity and trade aspects*

The Japanese Ministry of Agriculture has attempted to quantify the prospects for agriculture up to 1982 in the form of projections for major commodities of demand and production, and of implicit self-sufficiency ratios and import requirements. While these projections do not necessarily reflect official government policy they have been used in government planning. A review of agricultural policy currently being undertaken by the government is unlikely to change significantly the estimates of demand but could do so with regard to the relative contributions of domestic production and imports to fulfil total consumption requirements.

In the context of international agricultural adjustment, the increase in imports which is foreseen is the most important aspect of the projections. Two questions arise: (1) the size of the imports and the scope for some flexibility in their composition or quantities, and (2) the role of developing countries as suppliers.

*Future imports.* Imports would increase (in terms of value at constant prices) by some 6 to 7 per cent per year and thus approximately double over the 1970–82 period. The over-all self-sufficiency ratio (in value terms) for the commodities projected would remain approximately 75 per cent, or virtually the same as in the early 1970s. If however account is taken of the large contribution from imported feeding stuffs, the self-sufficiency ratio would, in terms of 'original calories', decrease from 53 per cent in 1970 to around 46 per cent in 1982.

On a commodity basis, Japan would be self-sufficient in rice, vegetables and eggs and mostly self-sufficient (i.e. 80 to 95 per cent) in meat, milk and milk products and fruits. Imports of the latter products would, however, increase. There would be only a small degree of self-sufficiency in concentrated feedstuffs, wheat, barley, soybeans and sugar. Projections were not made for oilseeds, oils and fats other than soybeans, nor of agricultural products not produced within the country. From a trade point of view, Japan is thus envisaged as maintaining complete self-sufficiency in rice, which still provides over one-third of the total food calories, while its agriculture becomes much more a converter of imported feeding stuffs into livestock products, and imports of other foods increase fairly slowly. The Japanese level of self-sufficiency of 70 to 75 per cent, or around 50 per cent if imported intermediate products are excluded, is certainly well below the average of almost all other industrialised countries.

Technically, it would seem possible for Japan either to raise or to lower its level of self-sufficiency as compared with the further small decline envisaged in the projections. To raise the level is an objective which commands some support in the country for reasons of food security, especially following the supply shortages of 1973–5 on world markets. To do so would, however, raise issues concerning higher consumer food prices or increased consumer subsidies, the cost to the economy in terms of the best allocation of national resources, as well as the need to control imports more tightly. Even to achieve the production goals of the projections which already imply an acceleration of the trend of production, wide-ranging efforts to increase productivity and to secure substantial changes in the pattern of output are considered necessary. Raising the level of self-sufficiency might also give rise to some conflicts with social–environmental requirements of putting more land to recreational use in a densely populated country with high and rising incomes; the income elasticity of demand in Japan for rural leisure facilities may now be much higher than that for food.

To lower the level of self-sufficiency below the projected level would imply a speeding-up of the outflow of labour from farming. Although this might be desirable on the basis of economic criteria of national resources allocation and labour productivity, a sustained outflow of more than the 5 per cent per annum envisaged in the projections would be likely to raise difficult social issues. Because of the environmental problems already being encountered in the major urban areas, the acceptability of such an accelerated transfer would probably depend mostly on the speed at which additional non-farm employment could be created in villages and smaller towns.

The second major issue which would arise is whether it would be prudent for Japan to choose to increase her dependence on imports of food and feed. This essentially involves assessment of the assurance of availability of import supplies, not in the sense of year-to-year availability which could be largely managed by national stock arrangements, but in the sense of longer-term availability. From this point of view, a Japanese policy which aimed at preventing the food self-sufficiency ratio from falling much further is understandable, especially when the maintenance of present self-sufficiency will lead to an increase in imports of 6 to 7 per cent annually. Given adequate national stocks, however, consideration might be given to at least a greater degree of dependence on imports for certain commodities, for example rice. A 95 per cent level of self-sufficiency in rice, in place of the envisaged 100 per cent, would lead to imports of more than half a million tons per year. Maintenance of 100 per cent self-sufficiency in rice, the basic staple in the Japanese diet, may however tend to make increased dependence on imports of other food items more acceptable.

*Role of developing countries.* The share of Japan's agricultural imports provided by developing countries fell from 42 per cent in 1960 to 36 per cent in 1971. Reduced imports from Asia and the Far East accounted entirely for the decline. Although the increase in Japanese import requirements between 1970 and 1982 will be very large, the competitive position of exports from developed countries for such products as grains, some oilseeds and livestock products is likely to remain relatively strong. Thus, the share of the developing countries in Japan's agricultural imports may continue to decline unless appropriate counter-measures are taken.

To maintain or to increase their share of Japan's large and rapidly expanding markets, supplies from developing countries will need to be competitive with those from developed countries in cost and quality as well as to be available in large quantities of uniform quality. Thus, it will be advisable for developing countries to expand or develop new production for export of some of these commodities, as is feasible. Fortunately, in order to secure a stable supply in sufficient quantities, Japan is seeking to diversify her sources of agricultural imports and therefore appears ready to provide developing countries at their request with technical and economic cooperation in this regard. It may prove desirable, however, for this assistance to be complemented by some sort of preferential arrangements in Japan for imports from developing countries.

# II Case Study of Poland

## CONTENTS

# 4 Long-term Changes in Polish Agriculture

## THE HISTORICAL SETTING

The structure of Polish agriculture before the Second World War was highly unfavourable. More than 47 per cent of agricultural land was in the hands of large landowners, and 65 per cent of the rural population lived on 15 per cent of the agricultural land. A relatively large social group consisted of landless rural families who depended for employment on the big estates or farms of the wealthy. Out of 3.2 million individual farms, 1.9 million had less than five hectares of land. The owners of these small farms, together with one million landless rural families and agricultural workers, were largely dependent on the some 15,000 large land-owning families and on about 200,000 well-to-do farmers.

The main obstacles to development of agriculture were the considerable feudal remnants, limited investment and consequent under-employment. The low level of industrialization and the slow rate of economic development hampered the outflow of existing large surpluses of man-power from rural to urban districts, with the result that agrarian underemployment was increasing and wages declining. It was estimated then that about 8 million persons, i.e. one-third of the total rural population, could leave agriculture without lowering agricultural production. In this situation labour productivity in agriculture was very low and showed no upward trend.

Agricultural production during the entire inter-war period of twenty years was largely stagnant, and crop output per inhabitant was no higher in 1938–9 than in 1909–13. Agriculture was marked by considerable technical backwardness. The number of tractors did not exceed one thousand on peasant farms; there were two binders and harvesters and four sowing machines per 100 hectares of agricultural land. Only 3 per cent of the villages were electrified, and the consumption of fertilizers amounted to only 6.7 kg of NPK per hectare of agricultural land, many times lower than in already advanced European countries.

The war worsened this condition. Almost 470,000 peasant farms were destroyed or damaged. The number of horses declined by 45 per cent, cattle by 60 per cent, pigs by 72 per cent and sheep by 63 per cent. About 20 per cent of the houses and farm buildings as well as about 30 per cent of the production installations fell into ruin. During the war one-fifth of Poland's population perished, mainly men at the age of professional activity.

Consequently, the level from which agricultural reconstruction and development started in Poland in 1944/45 was extremely low and points up the success which has been achieved in the past thirty years. It seems that this experience could in some respects also be relevant for other countries entering the period of accelerated economic growth. Particularly interesting is the fact that individual farming remains very important and in 1974 accounted for 80 per cent of the agricultural land and provided 77 per cent of the agricultural commodity production.

## LAND REFORM AND STRUCTURAL CHANGES

The process of thorough socio-economic change in the rural areas of Poland was initiated by the land reform proclaimed on 6 September 1944. This decree was one of the first and basic legal acts of the new government. The essential purpose was to abolish the system of feudal ownership and to transfer the land to direct use by peasants or to state ownership. Thus, the land reform simultaneously encompassed economic and socio-political aims. The farms subject to land reform were all of those above 50 hectares of agricultural land situated in the central and eastern voivodships,[1] and those above 100 hectares in over-all area in the western and northern regions. The farms of persons who had left Poland were also taken over by the State.

On the basis of the land reform decree the State Land Fund was established, which collected land and other means for land reform tasks. It was considered that newly formed farms or those enlarged by reform areas should not exceed 5 hectares of average-quality agricultural land. The decree banned division, sale or lease of farms set up in consequence of land reform. The charge for the land was symbolic and equal to the value of the average yearly rye yield of one hectare. This was five to seven times lower than the price for the land sold before the war under the land reform being carried out by the government of that time.

People who benefited from the land reform were landless rural families, agricultural workers, small tenants and owners of small farms. As a result of the reform and settlement, 814,000 new farms were set up and 254,000 farms enlarged. In the years 1945–9, more than 6.1 million hectares of land were allocated for these purposes (about 30 per cent of the agricultural land of the country). Almost 1.1 million farm families benefited by the reform and the area of small farms was enlarged on average by 1.9 hectares.

Owing to the land reform, the pre-war tendency towards polarization of Polish farms into two distinct categories in terms of land area – the large and the very small – was slowed down and reversed. This was reflected in the decline of the number of the smallest farms (up to 2 ha.) by 10 per cent and also in the number of the largest ones (above 20 ha.) by more than two-thirds and in the increase in the number of farms within the range of 2–20 hectares. The group of landless population ceased to exist, and a considerable number of former agricultural workers became individual farmers.

One of the most far-sighted decisions made in connection with the land reform was the setting up of the State sector in agriculture. Because of this, land reform became, in Poland as in other socialist countries, the first phase of socialist change in agriculture. As early as 1948 State farms covered about 8 per cent of the agricultural land. The land reform, while abolishing the remnants of feudalism, consolidated small-scale private farming and at the same time created the beginnings of a socialist sector.

## RECONSTRUCTION OF AGRICULTURE IN THE IMMEDIATE POST-WAR PERIOD

In the years 1946–9 the rebuilding of agriculture and the solution of national food problems were given a high priority. In 1946 a three-year plan (1947–9) of the country's economic reconstruction was prepared. In agriculture it provided for a rise of more than 10 per cent in per caput production compared with pre-war figures. This was to be based on the cultivation of wastelands, renewal of livestock populations, increase of crop yields and animal productivity as well as the increase in the supply of agricultural production means of industrial origin. In spite of the tremendous difficulties caused by war destruction, agriculture quickly achieved considerable results, primarily by the cultivation of wastelands and the launching of production in the western territories. Self-sufficiency in food production was not, however, achieved in the first post-war years. The targets of the three-year plan for crop production were accomplished fairly easily, but the losses in animal production were much more difficult to make up and took many years. Owing to increased government investment outlays, to credits, to electrification of rural districts and to promotion of education and agricultural training, the production tasks of the three-year plan were generally fulfilled by agriculture.

The new shape of the socialist sector in agriculture came into view immediately in the first years of economic reconstruction. State farms set up following the land reform developed and became consolidated. Also cooperative organizations were being created. The social and professional peasant movement became active. The Peasant Self-Help Union was established, a common cooperative organization of farmers with the task of extending agricultural skills. Many initiatives were taken also in the organization of production. In the years 1947–9 over three thousand cooperative machine centres were organized, which partly undertook to provide production services to individual agriculture. This was of essential significance, particularly for smallholders, who were unable to buy on their own account machinery and agricultural equipment. Although the number of these machine centres was not large, they nevertheless constituted the initial new form of cooperative enterprises in rural areas and they paved the way for technical progress.

At the same time other forms of cooperative activity were also developing which had a long-standing tradition in Poland, particularly cooperatives of sale and supply (still today servicing rural areas), credit cooperatives, dairy, horticultural and others. In those years there also came into being a new form of production link between individual farming and the State and cooperative organizations, namely, the contract-buying of agricultural products. This soon became the main organizational form of the agricultural market, of shaping planned production and of influencing individual farming by the State.

## AGRICULTURE IN A RAPIDLY INDUSTRIALIZING ECONOMY

The period of post-war economic reconstruction in Poland was followed by industrialization. The six-year plan (1950–5) was introduced. This was a very special period in the development of the country and of its agriculture. By changing essentially the economic strategy, it probably exercised the greatest influence so far on the structure of the Polish economy and it met with many new and difficult problems. The six-year plan assumed primarily a rapid rate of growth of the industrial potential, a quick development of building and a considerable outflow of population from agriculture to other activities. First of all, there was an expansion of investment in basic industries, which was bound to influence unfavourably the scale and the rate of growth of investment in agriculture and, consequently, the rate of growth of agricultural production. During the first phase of industrialization, the main effort was aimed at the construction and development of those industrial sectors which increased the country's economic potential and allowed Poland to overcome in a fairly short time its backwardness. This meant concentrating investments in mining, metallurgy, machine and chemical industries. Investments for the development of light industry, the food industry and agriculture were given second place.

The assumption of the strategy of industrial development sharply emphasized the problem of capital formation and of creating economic surpluses for the purposes of general economic development. The capital accumulation of industry itself, though constantly growing, was insufficient for financing the general indistrialization programme. Agriculture, as one of the major sectors of the national economy and providing 35–40 per cent of the national income, had to constitute a basic source for accumulating capital for economic development purposes, particularly for industrialization. For Poland in the early fifties, this was a necessary choice. The creation of this surplus, its concentration in the hands of the State and its productive use provided the crucial points of the economic strategy of those years. The contribution of agriculture to the general economic development in Poland thus took two main forms, namely the transfer of labour resources and the accumulation of capital means.

*1 The contribution of agriculture through the transfer of labour*

Agriculture has been and still is the major reservoir of manpower for expanding Polish industry. In Poland's post-war advance, particularly at the time of the country's rapid industrialization, the population migration from rural to urban districts played a tremendous role. In the first post-war years the growth of industry was of an extensive nature and, consequently, the production increase occurred first of all as a result of employment growth, and only to a smaller degree as a consequence of the increase in labour productivity. It is estimated that at the early stage of development over the period 1946–50 about 70 per cent of production growth was achieved by the increase of employment and only 30 per cent by higher labour productivity.

The population migration from rural to urban districts brought a double benefit to the economy. Those migrating from rural districts were, primarily, persons who were under-employed in agriculture and whose labour productivity there was very low. Quite often their marginal labour productivity was lower than their living costs. In this situation transfer of this manpower from a sector of low labour productivity (agriculture) to a sector of higher productivity (industry) benefited the national economy. It also brought about favourable results in agriculture itself, since the remaining manpower could be used more productively to increase output.

The main form of population outflow from the rural districts in post-war Poland was migration for permanent settlement. It is estimated that in the past thirty years such migration from rural districts totalled about 5.1 million persons with more than one-third of this outflow (1,816,000 persons) in the decade 1950–60. That was a most essential contribution of agriculture to the development of industry.

Another source of rural manpower for the growing industry – and a specific form of migration – has been and still is the group of part-time farmers, the main part of whom are peasant workers. The scale of this phenomenon in Poland over the whole post-war period has remained considerable. For example, in 1957 it was estimated that among the majority of smallholdings at least one family member earned his living in the socialized sector, while in some 45 per cent of smallholdings the head of the family was earning income outside agriculture. In 1960 23 per cent of the total number of professionally active members of farming families were employed off the farm and ten years later this proportion had risen to 35 per cent. There are few countries in Europe where every third adult member of farming families, as in Poland, has two working places and two sources of income. It must be emphasized that the availability of a large group of part-time farmers was extremely advantageous for the total national economy, in particular at the initial stage of its accelerated growth. Since the peasant workers continued to reside in villages, the national economy was relieved from additional costly housing and social construction. It was true that the peasant workers commuting daily to work burdened public transport, but at the time of acute material shortages this variant was cheaper and easier for realization. It was not insignificant either that peasant worker families had their own food supply, which again relieved the pressure on the market for consumption goods and also relieved the retail trade, then poorly developed. The contribution of peasant workers to the advance of Poland's economy was very great.

The contribution of agriculture through migration to the total economic development was reflected in the following two streams:

(a) about 5.1 million persons who, during the past thirty-year period, moved permanently from rural to urban districts, providing manpower without which a rapid growth of the non-agricultural sectors of the national economy would not have been feasible;
(b) from 1.6 million persons in 1950 to 2.8 million in 1970 who, continuing to reside in rural districts, earn their living outside agriculture; these now constitute 33.5 per cent of the total number of the professionally active rural population.

Thus, it is estimated that during the post-war period rural districts supplied the national economy with from 1.5 million man-years in 1950 to 3.2 million man-years in 1970 of fully efficient manpower from the combination of permanent migration and part-time farming. This is the approximate contribution through labour transfer or migration from rural districts and agriculture to the over-all economic growth.

*2 The contribution of agriculture through the accumulation of capital*

Another contribution of agriculture to the development of the national economy, particularly that of industry, was that of capital for investment. As noted earlier, particularly at the first phase of the accelerated growth, agriculture bears a part of industrialization costs. So it was also in Poland.

In order to ensure this contribution from agriculture to the efforts for industrialization, in the first years there was set up a system of compulsory deliveries of major agricultural products (cereals, potatoes, livestock and milk) at stable prices, much lower than market prices and lower than government-paid prices for the so-called non-compulsory purchases. It was in effect a tax in kind to be used for the accelerated industrialization of the country. In the first years the share of compulsory deliveries in the total production of agriculture amounted to about 20 per cent of cereal production, 10 per cent of potato production, 25 per cent of livestock production and 10 per cent of milk production. Prices paid for the articles covered by compulsory deliveries were within a range averaging about 50 per cent of the prices paid for the same products in contract-buying or in the free market. Although these prices covered the material costs of production they did not fully compensate for the farmer's labour. This part of the farmer's labour therefore became his contribution to the accumulation of social capital. Farmers could sell the remaining part of their production at higher prices, closer to the value of the products. The system of compulsory deliveries, which underwent several modifications, operated in Poland until 1971, when it was completely abolished.

Through the system of compulsory deliveries the State was solving two problems simultaneously. First it accumulated a surplus in the form of capital needed for the general economic growth, and secondly it secured for itself full availability of a substantial part of the agricultural production, which was at that time of essential importance for the urban food supply. At the same time, the system counteracted excessive and speculative rises in food prices in the conditions of market imbalance. However, the system of compulsory deliveries had a number of inherent drawbacks. It lowered the profitability of agricultural production and reduced the capacity of individual farms to save and invest. It also made the structure of agricultural production more rigid and weakened the incentives for increased output.

The system of compulsory deliveries was the main but not the only form of agriculture's share in meeting the costs of the country's industrialization. Another contribution was in the form of reduced agricultural investments. The investment programme in agriculture was very limited in those years, so that a part of the investment goods which would otherwise have been absorbed by agriculture could be used for the development of other sectors of the national economy.

The contribution of agriculture to Poland's industrialization and general economic growth, through transfer of labour and through capital formation, have been discussed in some detail because they have provided a major and effective element of Polish develepment strategy. They have allowed the State to concentrate considerable material and manpower resources and to use them for strategically important purposes. The development of industry became, in later years, a firm basis for the growth of agriculture and its technical modernization. Poland's experience clearly shows how the ability to concentrate the economic surplus in the hands of the State, however inconsiderable it may be initially, as well as the skill in using it, can provide a prerequisite for success of the whole strategy of accelerated growth. It has indeed been at the heart of the development process of the country. Agriculture can and must play a most essential role in the implementation of this strategy. This is particularly important in the countries where the share of agriculture in the national economy is relatively large.

## POSITION AND ROLE OF AGRICULTURE IN THE EXPANDING POLISH ECONOMY

Following the examination of the contribution of agriculture to economic development through migration of labour and accumulation of capital, it is appropriate to review the over-all position and role of agriculture in the Polish economy. Poland achieved a high rate of economic growth based on far-reaching socio-economic and structural changes after the Second World War. The national income, which amounted to about US $200 per caput before the war and did not exceed US $100 in the years 1944–5, reached US $1750 in 1974 at the official exchange rate. Agriculture, which was the main source of income for about 60 per cent of the population before the war, gave employment to only 27 per cent in 1973.

The primary factor bringing about the economic development in Poland and shaping the structural changes was industrialization. In comparison with 1938, total industrial production has grown twentyfold, with a nearly fivefold increase in the number of employees. Each employee in this sector is therefore now producing four times as much as in the inter-war period. In the post-war years there was created in Poland a powerful industrial base which facilitated the growth of all sectors of the national economy. It was not an easy task. By a tremendous investment effort the country was rapidly rebuilt after the war destruction, and its economic potential was developed and modernized in all fields of industry and agriculture, transport and communications, housing construction, and municipal and social and cultural undertakings. As a result of this effort the value of the fixed assets, amounting to 1.9 billion zlotys in 1946 (in constant 1971 prices), exceeded 4.9 billion zlotys in 1974. Traditional Polish industries such as fuel and power, textile and feed industries were developed and modernized. Poland changed from an agricultural–industrial country into an industrial–agricultural one with a large and modern production base, providing grounds for rapid development in the future.

The consequences and implication of industrialization in Poland have far exceeded industry itself. Industrialization becomes probably the prior condition for technical modernization of agriculture, which is followed by socio-economic reconstruction of agriculture. Without a powerful base of heavy industry it would have been unthinkable in Poland to increase very substantially the production of tractors and agricultural machines, fertilizers and pesticides, basic elements of the technical infrastructure of agriculture. The main factors which dynamize agricultural production and modernize the sector are nowadays outside agriculture, primarily in industry. Agriculture in the industrial era is unable to develop in isolation – it depends to an ever larger degree on industry.

The dilemma of Poland's economic policy was not whether to develop either industry or agriculture. The problem consisted rather in the correct determination of appropriate balances in the relationship between the advance of industry and that of agriculture. Although the economic policy did not avoid mistakes, it is beyond doubt that the development strategy which was selected proved to be right. While evaluating the industrialization process of Poland in a historical perspective, it should be emphasized that Polish experience provides further conclusions of an international significance. The success of the strategy of growth which was adopted provides a most convincing proof that medium-sized countries, retarded in their economic development, can overcome this underdevelopment largely by themselves but with some assistance in key areas. It is a conclusion which weakens frequently-held views, both in theory and in economic practice, which underestimate the possibility of developing countries overcoming their present economic backwardness. Of course, this is not an automatic process. It can be accomplished only if certain conditions are fulfilled, and in Poland the most important of these are thought to be the following: establishment of progressive socio-political relations, ensuring the guiding role of the State, particularly in accumulating economic surpluses and distributing them; and central economic planning and provision of social mechanisms allowing output capacities of the whole nation to be mobilized. For Poland the economic assistance of the USSR with favourable credit conditions was of basic significance. The USSR supplied Poland with designs and equipment and rendered technical help in building more than a hundred key industrial plants constructed after 1950. While this assistance was a major element in Poland's industrialization, the major source of the success of its industrialization policy was that the country solved the essential problems common to all countries retarded in their economic growth, namely the structural disproportion between labour and capital resources, by using social and economic mechanisms available in the conditions of a centrally planned economy.

The industrial development of the country did not overshadow the importance of agriculture. Though its share in the material production and in the national income keeps declining (from 55.8 per cent in 1947 to 11.8 per cent in 1974), its significance for a rapid and steady development of the country is great. Basic statistical data on the development of Polish agriculture and its place in national economy are given in Table 4.1. It should be noted in particular that agriculture meets directly or indirectly about 40 per cent of all consumption needs of the population, i.e. in 1973 40.8 per cent of all personal incomes were spent on food. It is characteristic that the average yearly rate of growth of expenditure on food was high (3.7

TABLE 4.1 Basic data on the socio-economic development of Poland's economy and of Polish agriculture, 1950–74

| *Item* | *Unit* | *1950* | *1955* | *1960* | *1965* | *1970* | *1974* |
|---|---|---|---|---|---|---|---|
| Population total | million | 25.0 | 27.6 | 29.8 | 31.6 | 32.7 | 33.7 |
| Urban | million | 9.2 | 12.1 | 14.4 | 15.7 | 17.1 | 18.2 |
| Rural | million | 15.8 | 15.5 | 15.4 | 15.9 | 15.6 | 15.4 |
| Share of non-agricultural population | per cent | 52.9 | — | 61.6 | — | 70.2 | 72.9 |
| Professionally active persons in national economy | million | 10.2 | — | 12.4 | 13.5 | 15.2 | 16.8 |
| National income (constant prices) | 1950 = 100 | 100 | 151 | 208 | 280 | 374 | 548 |
| Structure of national income | per cent | 100 | 100 | 100 | 100 | 100 | 100 |
| Industry | per cent | 31.8 | 36.3 | 39.5 | 45.0 | 49.8 | 51.3 |
| Agriculture | per cent | 47.9 | 32.7 | 27.3 | 22.6 | 15.8 | 11.8 |
| Consumption | 1950 = 100 | 100 | 149 | 204 | 260 | 340 | 475 |
| Accumulation (savings) | 1950 = 100 | 100 | 164 | 243 | 363 | 499 | 1074 |
| Investment | 1950 = 100 | 100 | 177 | 274 | 410 | 635 | 1367 |
| Share in national income | per cent | 21.0 | 22.5 | 24.2 | 26.8 | 27.9 | 37.3 |
| Total industrial production | 1950 = 100 | 100 | 200 | 317 | 474 | 708 | 1048 |
| Total agricultural production | 1950 = 100 | 100 | 110 | 132 | 152 | 166 | 195 |
| Average yields | | | | | | | |
| Grains | q/ha. | 12.7 | 14.3 | 16.1 | 19.2 | 19.6 | 28.3 |
| Potatoes | q/ha. | 138 | 100 | 132 | 154 | 184 | 181 |
| Sugar beet | qw/ha. | 222 | 186 | 256 | 259 | 312 | 295 |
| Animals per 100 ha. of agricultural land | | | | | | | |
| Cattle | actual | 35.2 | 38.8 | 42.6 | 50.7 | 55.5 | 67.6 |
| Pigs | actual | 45.7 | 53.4 | 61.8 | 70.2 | 68.8 | 111.6 |
| Meat production | thousand tons | 1323 | 1378 | 1356 | 2015 | 2187 | 3072 |
| Milk production | billion litres | 7.8 | 9.6 | 12.1 | 12.9 | 14.5 | 16.3 |
| Agricultural tractors (excluding hortic.) /in 15 HP units | thousand units | 25.5 | 59.1 | 77.2 | 154.3 | 278.9 | 475.6 |
| Use of fertilizers (pure component) | kg/ha. agric. land | 17.6 | 25.8 | 36.5 | 56.4 | 123.6 | 173.6 |

— indicates figures not available.

per cent in 1966–70, 5.1 per cent in 1971–3). The income elasticity of food demand remains fairly high in Poland and is estimated at 0.537. The coefficient of price elasticity for food is estimated at −0.448. Thus the significance of agriculture for feeding the country's population is greater than it would seem by its share in national income. The demand for highly processed food articles and for those containing protein of animal origin is growing dynamically. These trends influence considerably the structure and dynamics of growth of agricultural production.

The food problem has been basically solved in Poland by increased agricultural production. During all of the post-war period Poland has been self-sufficient in food. There have been continuing imports of certain quantities of grains, raw materials of agricultural origin, tropical fruits, condiments and beverages such as tea, coffee, etc. and other food articles. On the other hand there have been exported considerable amounts of various agricultural products (cattle on the hoof, meat and meat products, some fresh and processed fruits, alcohols, etc.) so that the foreign trade balance in agricultural products was favourable.

The level of nutrition in Poland is fairly good. According to 1972 data the energy value of an average daily diet amounted to 3360 calories, including the following (in grams): protein, 91.4 (including animal protein, 49.8) fats, 110.4, carbohydrates, 502.0. In the most recent years supplies on the market in Poland have been adequate to meet demand, thus satisfying the essential calorie requirements. Much progress in this respect was made after 1970 and, which is of particular significance, it was achieved with a freezing of retail prices for basic food articles and with simultaneous growth of per caput incomes (in 1970–4 the real wages for work and the social benefits increased by 57 per cent, or 36 per cent per caput). The per caput consumption of meat and edible animal fats has grown from 61.2 kg in 1970 to 73.8 kg (including 65.5 kg of meat) in 1974. The consumption of other products per caput in 1974 is as follows: cereals and cereal products, 123 kg, potatoes, 177 kg, fish and fish products, 7.3 kg, milk, 414 litres, eggs, 205, sugar, 44 kg.

Poland's economic development in the whole post-war period was carried out under the strong influence of an 'agricultural barrier' to growth. A characteristic feature of all countries at a level of economic development similar to that of Poland is that the attainable increase in agricultural production limits the rate of growth of incomes and living standards because, as noted above, the share of expenditure for food is relatively high. Under these circumstances the rise of wages and of incomes in real terms depends, to a large degree, on the supply of products of agricultural origin. According to Polish experience, in these conditions an increase in the share of non-food expenses is particularly desirable, as it lessens the pressure of demand on agricultural production and renders the increase of actual wages more independent of the current production capacities of agriculture.

The position of agriculture in Poland's national economy cannot of course be limited to the problems of food. Agriculture not only produces food but also presents definite demands for production means, services and consumption goods of industrial origin. As a consequence of the economic growth and expanding division of labour, agriculture is ever more strongly linked with and dependent on the national economy. Agriculture appears to absorb 7.3 per cent of the value of industrial products and services, 27.2 per cent of the value of buildings and constructions and 8.2 per cent of the value of total material goods and services produced by non-agricultural sectors. At the same time, about one-third of the production means used in agriculture comes from outside that sector. These are primarily products of the chemical industry (fertilizers and pesticides), the electrical and machinery industry, the fuel and power industry and the food industry. Simultaneously, the market orientation of agriculture is growing and its demand is increasing both for industrial production means, such as agricultural machines and equipment, fertilizers, manufactured feeds, power sources, building materials, etc., and for processed consumption goods. Thus agriculture is becoming less self-sufficient and more fully integrated into the national economy.

In spite of a high rate of expansion, the inflow to farming of production means and consumption goods from non-agricultural sectors has not been sufficient so far, a fact which has hampered and slowed down the process of technical reconstruction of agriculture and its modernization. The development plans for the next two decades assume investment in means of production in agriculture more than treble the level of comparable investment in the whole post-war period. This will provide conditions for a substantial technical modernization of agriculture.

Changes in the character of agricultural production are also shown in the structure of the distribution of its output. According to data on inter-sectoral flows, 33.3 per cent of the value of total agricultural production in 1973 went to industry, 22.3 per cent to animal product processing and 8.8 per cent to plant processing industry. A large part of agricultural production (42.8 per cent) remained in agriculture, and, in the form of fodder, seeds, etc., was available as an input for further production. Thus in all 76.7 per cent of the total agricultural product served in 1973 to satisfy the so-called indirect demand, namely the requirements of material production. The remaining part, i.e. 23.3 per cent, was allocated for satisfying final demand, 16.8 per cent of which was consumed directly by the population and 2.2 per cent designated for exports. Studies of inter-sectoral flows show that Polish agriculture produces a rising proportion of production means and raw materials for food processing (satisfying the demand of the material production sphere) and relatively less food for direct consumption (satisfying the final demand). This confirms the thesis that in the course of economic advance, agriculture becomes to an ever larger degree a raw material branch with the production of final consumption products undertaken by the food processing and distribution industry.

## MANAGEMENT OF AGRICULTURE DURING INTENSIVE INDUSTRIALIZATION

The increased investment effort for setting up an industrial basis for economic development was bound to affect unfavourably the growth of agriculture and of agricultural production. In the first years of the six-year plan (1950–5) agricultural production was growing at the average rate of 2.5 per cent a year, which, considering the marked population growth and a significant migration from rural to urban districts, brought about temporary food shortages. In the years 1951–5 the value of production of individual farms increased in comparison with 1950 by only 5 per cent, and in the whole six-year period at only 1.8 per cent a year. Thus, the consequence of reduced outlays for agriculture was a marked drop in the rate of production growth which led to an imbalance between the growth of demand for agricultural products and the growth of supply. This phenomenon appeared acutely in the form of temporary food shortages on the home market, particularly shortages in animal products, and in the form of a deteriorating foreign trade balance in the sector of agricultural food products.

The lowering of the rate of growth of agricultural production at the time of accelerated industrialization was the price which had to be paid for excessive – as it is now evaluated – burdening of agriculture with the costs of development of the national economy. The experience of those days indicates that, in charging rural districts and agriculture with part of the industrialization costs, it is necessary that the burden should not be excessive. It brings about a decline of farmers' activity, lowers the rate of growth of production, increases food problems of the country, and, if this state of affairs lasts for long, it causes a decline of the capacity of individual farms to produce a surplus, which is a source of investment funds.

Although there was only limited growth in output in the years 1950–5, many favourable changes took place in agriculture during this period, in particular a wider access to work for the rural population, abolition of the remnants of feudalism, and improved prospects for rural youth by widespread training and by preferential treatment for this youth in admission to schools. The difficulties mentioned above were also aggravated in those years by attempts at rapid collectivization of agriculture. These attempts were undertaken in Poland in the years 1949–50. In the first period they did not give good results. Polish rural populations were not ready at that time to accept the collective form of farming, both for material reasons (lack of modern production techniques, e.g. there were produced then in Poland only 5–7,000 tractors a year) and for psychological reasons. Consequently there were attempts through administrative pressure for setting up production cooperatives, which caused the formation of bodies unable to act independently or to develop rapidly. In the years 1950–5, 10,510 production cooperatives were organized. These were mostly economically weak units, not adapted to the operation of modern, large-scale farming, and they did not stand the test of time. Most were dissolved when the course of agrarian policy towards individual farming changed. In 1956 there remained in Poland only 1534 cooperatives. Only the strongest units survived. These were consolidated afterwards and gave rise to the cooperative sector in Polish agriculture.

It is worth noting that in 1955, when the greatest number of production cooperatives was operating, their potential was not great. They associated 188,000 families (including only 165,000 families owning farms) and their area accounted for only 9.2 per cent of the total agricultural land. After 7542 cooperatives were dissolved in 1956, their share in agricultural land dropped to 1.2 per cent. Under these circumstances the economic importance of cooperatives was not great. Nevertheless, they performed vital social functions as regards the political system. When well organized, well managed and highly profitable, cooperatives provided an example to be followed by individual farmers who, though not in great numbers, joined the already operating cooperatives and set up new ones.

## NEW GUIDELINES OF AGRARIAN POLICY AND THEIR IMPLEMENTATION

Growing discrepancies between the increasing demand for agricultural products and their supply led to a revision of economic policy principles in 1954. The revision aimed at increasing outlays for sectors which produced consumption means, including agriculture. Decisions were made to raise investment outlays for agriculture, to increase the production of agricultural machines, equipment and fertilizers, and to improve the organization of the mechanization of agriculture. Measures taken up in 1954 did not at first give the expected results. The year which brought fundamental changes was 1956, with revisions in the guiding principles affecting both economic and agricultural policies.

By and large, these changes consisted of departures from administrative methods of agricultural management, and a switch to the broad application of economic instruments. With these changes, the new guidelines of agrarian policy were accepted and brought rapid results. The guidelines envisaged in particular the following actions:

(a) considerable increase in the supply to agriculture of industrial production means, primarily fertilizers, tools and agricultural machinery and building materials. Consequential shifts were made in the investment programme of industry in favour of sectors servicing agriculture;

(b) improvement of the profitability of individual farms for which purpose compulsory deliveries were reduced and State purchasing prices for agricultural products were raised. Economic relationships between agriculture and the total national economy were based mainly on contract-buying;

(c) stimulation of animal production through increased imports of cereals and fodder and through reduced cereal purchases from farmers, so as to increase the degree of processing of fodder into meat;

(d) settlement of the matters connected with land ownership;

(e) reactivation of many of the self-managed peasant organizations which had been dissolved in the years 1949–56, and especially the establishment of agricultural circles and activation of various forms of rural cooperatives (cooperatives of sale and supply, credits and loans, dairy, horticultural and other cooperatives);

(f) introduction of organization principles for State farms, aiming at the self-dependence of single farms, and introduction of a self-financing system as a stimulus for productivity growth.

It should be emphasized that these changes coincided with the process of democratization of social life in the country and with the creation of a new climate regarding individual farming. Individual farms were acknowledged as a long-lasting element of the agrarian structure and, at the same time, conditions for their development were created. In these ways an activation of farmers and of their self-management and of social-professional organizations was achieved. Interest in agricultural production and its increase grew. The rate of growth of agricultural production increased from the average of 1.8 per cent a year in the period 1950–6 to 3.1 per cent in 1957–60, and in the subsequent five-year period 1961–5 to an average of 3.7 per cent a year. In the sixties there could already be made available to agriculture larger supplies of production means and fertilizers, designated first for industrial crops and contract-bought crops, which brought a considerable increase of yields. In addition to meeting domestic demand, increased production provided some surpluses for export such as sugar beet, rapeseed, grass seeds, etc. At the time of shortage of fertilizers a policy of concentrating their use on the fertilizer-responsive crops was introduced. Contract-buying served the implementation of this policy since a farmer who signed a contract was entitled to purchase a defined amount of fertilizers. Later, with the growth of domestic production of fertilizers, they became generally available and the special allocation policy could be ended. Balance on the fertilizer market was achieved in Poland in the years 1965–8.

In the years 1965–70 the rate of growth of crop production accelerated as a result of the widespread introduction of more rational seed management and modernization of the disease, pest and weed control in field crops and in fruit growing. Seed production played a great role. Many new varieties of cereals and other plants which were developed by Polish scientists and growers were introduced into cultivation.

### AGRICULTURAL DEVELOPMENT FUND AND THE AGRICULTURAL CIRCLES

As a consequence of political decisions in 1956, agricultural circles were reactivated in Poland. These are widespread peasant organizations of a social – production character, having a rich tradition and experience. Attention is given here to them since the agricultural circles are an original solution, typical for Poland, and because they play an essential role in the modernization of individual farming and in expanding its socialization.

The activities of an agricultural circle cover, in principle, one village. It is an organization set up on a voluntary basis which has the authority of an agricultural self-government as regards the entire village. The main task of the circles is the introduction of progress in agricultural production. They are fulfilling these tasks by developing a wide activity in the field of production services to agriculture (mechanization of land cultivation and harvesting, plant protection, investment, building and transport services). In relation to farmers of the given village the agricultural circle is first of all a service organization. It usually has available a considerable social property, mainly machines and equipment for agricultural production, and renders services to farmers against payment.

The basis for the activity of the agricultural circles is the Agricultural Development Fund set up in 1959. Beginning with 1960 the State allotted to the village (and not to single farmers) the sums resulting from differences between the prices for compulsory deliveries of agricultural products and those paid by the State in free market purchases. Prior to 1960, compulsory deliveries were a contribution of agriculture to the development of the national economy. With the creation of the Agricultural Development Fund these resources remained in the village to provide a social fund for the mechanization of agriculture. Since the abolition of compulsory deliveries in 1971, means for this fund come from taxation of individual farms. These means are not transferred to individual farmers but to their socio-economic organization, i.e. to the agricultural circle. The Agricultural Development Fund can be designated only for purposes planned ahead, primarily for the purchase of major agricultural machines for common use. In this way there is created a social property of production means for individual farming, without socializing the land.

The Agricultural Development Fund and the agricultural circles operating under its support provided the institutional and organizational framework for the technical reconstruction of Polish agriculture. At the same time, they formed material – technical conditions for its social reconstruction. This is the process, specific to Poland's conditions, for adjustment of individual farming and small-scale farming to the requirements of modern technology. It is considered in Poland that in the conditions of that country an individual farm, as a form of production organization and as the source of income of a farmer's family, has undoubtedly become outdated. It is unable to secure for itself access to modern techniques by any way other than group mechanization. Agricultural circles are meeting the problem half-way by providing a material base of technical service for individual farming. In 1970 the agricultural circles and their unions had available property totalling 32 billion zlotys, which constituted 4.1 per cent of total fixed assets in agriculture, 5.1 per cent of fixed assets in individual farming and 11.8 per cent of fixed assets operating in the socialized sector of agriculture. The agricultural circles also had at their disposal about 35 per cent of the total number of tractors in the country and a considerable number of other heavy machines and transportation means.

The activity of agricultural circles is however not limited to common use of agricultural machines and to rendering services to individual farms. Some of the circles operate group production of plants and animals on the land of the State Land Fund (government-owned land) or in village land communities. A certain number of agricultural circles also started consolidating land of their members in order to carry on common cultivation of cereals and other plants, common orchards or common vegetable production (e.g. hothouses), pig breeding, slaughter cattle breeding, poultry breeding, etc. These processes became intensified particularly after 1971.

## NEW STRATEGY AND MAIN DEVELOPMENT PROBLEMS OF AGRICULTURE AFTER 1970

The change of political leadership in December 1970 brought about an essential change of the strategy for the country's economic development. It was reflected, first of all, in the provision of strong economic incentives for the increase of labour productivity. The limitations on wage increases existing in production enterprises were cancelled and wages became conditional on increased labour productivity and on economic results of the enterprise operations. Priorities in economic development were changed. The main aim of development was acknowledged to be improvement of the material prosperity of the society and the satisfaction, as far as possible, of people's needs. It was accepted that providing a chance for steady and considerable wage increases and of satisfying growing needs in this way is one of the essential factors of economic growth, since it releases the initiative of individual workers and of whole collectives, promotes a better organization of work and increases its efficiency.

The result of the new economic policy has been a considerable acceleration of the country's economic growth, as shown in Table 4.2.

TABLE 4.2 Average yearly rate of growth of national income (at constant prices)

| *Years* | *Aggregate* (%) | *Per caput* |
|---|---|---|
| 1951–5 | 8.6 | 6.5 |
| 1956–60 | 6.6 | 4.9 |
| 1961–5 | 6.1 | 4.9 |
| 1966–70 | 6.0 | 5.2 |
| 1971–4 | 10.0 | 9.1 |

The accelerated development of the national economy in recent years was associated with simultaneous increases of wages and incomes of the population. In the years 1966–70 personal consumption expenditures per caput increased on the average by 4.3 per cent yearly, while in the years 1971–4 it rose to an average of 6.4 per cent yearly.

The fast growth of the national income allowed a higher rate of growth of people's incomes together with a decrease in the share of consumption expenditures in the national income. Total consumption expenditures as a proportion of the gross national product declined from 65 per cent in 1970 to 58 per cent in 1973, while the share of gross capital formation increased from 35 to 42 per cent. Consequently, the rate of growth of investment outlays rose considerably, the increase averaging 8.1 per cent yearly over 1966–70. Such a high rate of growth of investment outlays provides a basis for maintaining the present rate of economic growth in the near future. Also, agriculture had to become adjusted to the general changes in the socio-economic policy. The change of priorities in economic development, with the satisfaction of the material needs of the population becoming the primary objective, set great production tasks to agriculture.

The agrarian policy after 1970 implements three main aims. Production aims come definitely to the fore. The major issue became the increase of agricultural production up to the level which would ensure meeting the ever-growing demands of the home market, of the agricultural-food industry and of exports. The premise is that agricultural-food exports should at least cover the costs of imports of production means for agriculture, and of agricultural commodities for the domestic market which are not produced in the Polish climatic zone. The importance of exports of sophisticated agricultural products is greater in so far as they constitute a permanent item in the trade with some highly advanced countries. The increase of agricultural production is essential for a harmonious economic development, for ensuring social conditions of this development and for maintaining a permanent balance in the market. Many undertakings of the agrarian policy of recent years have been subjected to these production aims.

Another guiding aim of the agrarian policy is to provide conditions for gradual, structural changes in technical and social aspects of agriculture. It is being realized in Poland that the structure of agriculture is backward and does not suit the contemporary stage of scientific – technical revolution in agriculture, nor does it offer conditions for a rapid advance in the longer-term. Provisions are therefore made for the gradual concentration of land, comprehensive production cooperation and increased scale of production, as well as for the formation of new, socialist production relations.

The achievement of ambitious adjustment aims (an increase in agricultural production of 65 per cent in the 15 years to 1990 with a 50 per cent reduction in agricultural employment) as well as technical and social reconstruction of agriculture will not be possible to implement if social conditions of the rural population are not solved simultaneously.[2] Therefore, the third main target of the present agrarian policy is a considerable increase in individual incomes of the rural population, improvement of working conditions, introduction of health and social insurance for individual farmers and members of their families, facilitating access to cultural achievements for rural population – in brief, equalizing of living and working conditions in urban and rural districts. The policy premise is that the place of birth and of work should not be a reason for social and cultural discrimination. This is a long-range and very costly programme requiring consistent implementation.

An important aspect of the spirit and the sense of all political measures after 1970 is that while previously the agrarian policy of the State reacted, in most cases, to disturbances and incorrectnesses in development by *ex post* action, the basic orientation of policy is now to prevent the occurrence of such disturbances, i.e. to influence the development of agriculture by *ex ante* action. The present agrarian policy includes many anticipating elements which consciously shape the emerging reality. In particular, the role of economic instruments for directing and managing agriculture (prices, conditions of production, profitability, credit and other economic incentives) has been increased, thus limiting administrative measures to the necessary minimum. The basic premise of agricultural policy is that socially needed production must be profitable for every producer. While observing this rule, major changes were carried out in the price system for agricultural products. Prices were raised by almost 30 per cent for animal products, primarily meat and milk, and imports of grain and high-protein feeds were increased substantially. Thus, indispensable conditions for intensification of the farm fodder production, which would be profitable in relation to the price of the final product, were met. This fact has far-reaching implications as regards intensification of crop production.

While setting up conditions for modernization of agricultural production the following moves, in particular, were undertaken.

1 Beginning 1 January 1972 compulsory deliveries of livestock, cereals and potatoes were abolished and the range of contract-buying of agricultural products at higher, profitable prices was considerably enlarged. Consequently, the purchasing prices for agricultural products increased and the financial burdens of farmers declined. Prices for beef cattle, pigs and milk were raised on the average by about 30 per cent.
2 The Agricultural Development Fund was maintained as the main source of social capital formation in individual farming. This fund is now provided from the land tax while earlier it was linked with compulsory deliveries.
3 The tax system for individual farmers was changed, with the result that the economic situation of farmers cultivating poor soils was improved and the range of tax reductions for smallholdings was expanded.
4 Supplies of production means for agriculture, mainly agricultural machines and equipment, building materials, fuels and transportation means were increased.
5 Measures were undertaken to improve trade services to rural districts and agriculture, eliminating the waste of farmers' time when selling agricultural products. Investment outlays for the development of the agricultural-food industry and for technical services to agriculture were increased considerably.
6 Living and working conditions of individual farmers and of agricultural workers were improved by the introduction, among other things, of health insurance and by expanding the range of free medical care in rural districts.
7 Socio-political and material conditions were provided for development of farmers' self-government by consolidation of the importance and enlargement of the operation range of 'gmina' (the lowest cells of State authority).

Farmers clearly felt the improvement of economic production conditions. As a result of the general price increase for agricultural products and of profitability, changes surpassed all expectations. During the four years 1971–4 the agricultural production increased by 24 per cent (in constant prices), or by an average of 5.6 per cent yearly. This expansion was faster than assumed for the whole of the 5-year plan (1971–5). Such a rate of growth had never previously been achieved in Polish agriculture. Animal production in particular increased rapidly. Pig numbers increased from 13.5 million in 1970 to 21.5 million in 1974, i.e. by nearly 60 per cent. Cattle numbers increased from 10.8 to 13.0 million. Naturally, this rate of growth cannot be maintained for a long period of time. Nevertheless the large increase of agricultural production achieved in recent years means that implementation of the tasks envisaged for 1990 starts from a relatively advanced level.

The price decisions made and the rapid growth of commodity production in agriculture brought a considerable increase of incomes for rural population and raised its purchasing power, as shown in Table 4.3. Although total incomes of individual farms rose by 30 per cent during four years, the fiscal charges were not changed, resulting in their share in total income falling from 12.8 to 9.7 per cent. Thus, the income at the disposal of farms increased to a higher degree than the general income, namely by 35 per cent. The total consumption expenditure of the entire individual farming sector expanded at an average annual rate of 6.3 per cent, by 7.4 per cent per caput.

TABLE 4.3 Incomes of individual farms from agricultural production, 1970 and 1974 (constant 1971 prices)

| *Item* | *1970* (*billion zlotys*) | *1974* | *Increase 1970–4* (%) |
|---|---|---|---|
| Total income[a] | 120.5 | 157.3 | 30.4 |
| Fiscal charges | 15.4 | 15.3 | −0.7 |
| Agricultural income[b] | 105.1 | 142.0 | 35.1 |
| Investment expenditures | 1.3 | 9.4 | — |
| Including: | | | |
| Fixed capital formation | 9.7 | 15.8 | 62.9 |
| Increase of value of herd and stocks | −2.6 | 1.3 | — |
| Credit balance[c] | −5.8 | −7.7 | — |
| Consumption expenditure[d] | 103.8 | 132.6 | 27.8 |

[a] Value of final agricultural production less production expenses (without depreciation).
[b] Income remaining at the disposal of farms.
[c] Increase of farms' indebtedness.
[d] Together with expenses for housing investments.

Besides improvement of the economic situation of individual farms, the introduction in 1972 of social insurance for individual farmers was of fundamental significance. Individual farmers and their family members are now entitled to free medical treatment.

A characteristic feature of recent years is the rapidly growing demand of farmers for production means. This demand considerably exceeds the growing demand for consumption goods and thus clearly indicates a willingness of farmers to raise outlays for production. The interest of farmers in land utilization has also grown. A virtual revolution occurred in

Poland after 1965 in respect of the use of fertilizers, which rose from 56 kg NPK in 1965 to 174 kg NPK per hectare of agricultural land in 1974. Also the supply of agricultural machines increased considerably.

The agrarian policy is deliberately set on bringing into operation and exploiting all possible reserves and opportunities for the increase of agricultural production. Owing to investments and other governmental outlays, dispersed farmers' resources are mobilized, effective forms of social capital formation for production purposes are established and the activity and initiative of farmers themselves, as well as of their socio-professional organizations, are promoted. There are also some reasonably applied measures of administrative pressure, for example, in fighting plant pests, exchange of seeding material, post-improvement management of agricultural land, ban on calf slaughter, etc. These administrative measures are understood and find support from leading farmers because they are a factor in speeding progress and changing older practices by subjecting farmers at large to a definite agro-technical discipline, based on scientific grounds.

An ever-more essential factor of production growth in individual farms are services rendered by specialized, socialized enterprises (State cooperative and agricultural circles). Services, particularly in the mechanization of field works and transport, provide an important factor of production modernization in the traditional structure of individual farming and constitute a permanent element in farming changes. Greater resources of the most modern agricultural techniques and of highly qualified staff are provided through these services. This is, simultaneously, one of the ways of socializing agricultural processes.

A crucial element of the complex programme for production increase and intensification of agriculture is the land policy, understood as principles of rational land administration. There are in Poland 19.3 million hectares of agricultural land, including 19.1m hectares of arable land, 2.5m hectares of meadows and 1.7m hectares of pastures. Thus, there are 0.6 hectares of agricultural land of medium quality per inhabitant. There are no wastelands available in Poland to provide a land reserve for cultivation. Consequently, the only way to increase production is to intensify the utilization of existing land resources. The increase of land productivity is now a more important practical problem than is the increase of labour productivity, which is rising with the decline of agricultural employment and the increase of productivity of land and of livestock.

The agrarian policy attaches great importance to rational utilization of land. In Polish land policy the rule is followed that land, as a production means, is a national property and, regardless of the ownership form, must be fully used to the advantage of the whole society. Meanwhile, however, a part of the land is not used very intensively, since there are a certain number of economically weak farms lacking development conditions.

A basic problem of Polish agriculture is the aging of the rural population. Emigration to industry and other non-agricultural sectors is mostly by young people while elderly people remain in rural districts. Some farmers whose children have settled in towns have no successors. With the aging of those farmers, their land becomes progressively less effectively farmed. In 1974 farm operators over 60 years of age were 33.1 per cent of the total, this figure having risen from 31.2 per cent in 1967, and so a problem arises of taking over and rationally managing a part of the land. In order to counteract unfavourable results for agriculture and for the whole national economy, there was introduced in 1968 a bill to provide pensions for farmers transferring their land to the State. The bill enables a farmer who has no legal successor to obtain a life pension from the State in return for the land, transferred voluntarily. Land taken over by the State increases the State Land Fund, which has at its disposal 862,000 hectares of land at present.

Priority in the management of land taken over from farmers by the State against pensions goes to State farms and agricultural cooperatives. However, since the area of these farms is considerable, and the farms taken over by the State against pension are dispersed, it is often impossible to attach them to a State farm or to an agricultural cooperative. In such cases the land is leased by the State to individual farmers, who can guarantee to use it rationally. Thus, production criteria are determining this policy. It is in the interest of agriculture and of the entire society to transfer the land to those farms which can use it most effectively.

Transfer of land to the State, on the basis of voluntary decisions of farmers, leads to a gradual though slow transformation of private land ownership into socialist ownership. The scale of this process continues to increase.

Another procedure favouring rational use of land is land consolidation. In a major part of the country there occurs not only a considerable reduction in the number of farms but also a wide dispersion of lots (fields) within single farms, resulting in the so-called field chessboard which makes farming extremely difficult, raises its costs and sometimes rules out modernization of field work. In order to eliminate this drawback land consolidation on a large scale was undertaken in 1968, on the basis of a new law on consolidation of fragmented plots. This law simplifies and accelerates the process of consolidation and enables, at the same time, compact complexes to be formed out of the lands of the State Land Fund and of the farms transferred to the State against pension. In this way conditions were provided for gradually overcoming one of the factors slowing down and hampering intensification of agriculture.

Other elements of land policy are laws and operations aimed at counteracting loss of land from agricultural use. Regulations which make land turnover more flexible and systematize land management issues were also introduced recently. The target is, in particular, to facilitate free transfer of land from farms which are unable to use it to those which are able to do so. These transfers are effected either on the basis of a purchase – sale contract or, ever more frequently, of long-term leases.

A decisive role in long-range shaping of effective intensification of agriculture is played by investments. The leading rule of Polish agrarian policy is to associate the growth of agricultural production with the process of socialist change of the agriculture. The investment policy is subjected to this principle. It gives preference to social investment forms, and at present about 60 per cent of agricultural investment outlays are constituted by socialized investment. It means, in practice, a faster growth of socialist ownership of the means of production than of private ownership. These investments are financed by State funds, by the Agricultural Development Fund, by the funds of agricultural circles and by those accumulated in agricultural cooperatives. These investments serve both production growth in socialized agriculture and on individual farms. Individual farmers are also interested in the growth of investment of socialized enterprises, particularly of agricultural circles.

If the farmer can use at any time machines owned by the agricultural circle, he does not have to buy them himself. This helps him reduce his investment outlay, for a major part is taken over by the State or by self-governed farmers' organizations. From the social viewpoint the solution is effective since it decreases social costs of mechanization of agriculture and of exploitation of the expensive, specialized equipment.

The State policy towards agriculture is complex. It assumes both a considerable increase of requisites coming from non-agricultural sectors (agricultural machines, fertilizers, pesticides, building materials, fuels, electric power, etc.) and the mobilizing of farmers' own resources and stimulating their initiative and professional activity. The attitude of the farmers themselves, their skill, abilities and involvement decide, in the last resort, the effectiveness of all material undertakings initiated by the State. A special feature of the present policy is that its starting point in considering farmers is their economic interest. In sum, the assortment of economic instruments for influencing agriculture is very large.

## FOREIGN TRADE IN AGRICULTURAL PRODUCTS

For Poland, relations with foreign countries are an extremely important factor for rapid economic development. In view of the major share of agriculture in total production, agriculture takes high priority in foreign trade. Agricultural and food exports remain for Poland an important source for financing imports of industrial (mainly investment) products and are thus a prerequisite for national development. Although less than one-seventh of the total foreign earnings comes from agricultural and food exports, the significance of these exports for the country's economic development is much greater because of the geographical structure of these exports. Agricultural and food products provide over one-third of Polish exports to free-exchange markets. In recent years nearly half of the Polish exports to EEC countries were agricultural and food products, and earlier this share was even higher. Although the share of agricultural and food exports in the total Polish exports continues to decline, its actual volume is nevertheless growing. The traditional 'Polish' markets for meat, eggs and poultry, milk products, sugar and sugar products, fruit and vegetable products, alcohols, seeds, breeding animals, etc. are still a major source of foreign currency. As noted elsewhere, a general objective as regards agricultural trade is that exports and imports of agricultural commodities and production requisities should be maintained in approximate balance.

As shown in Table 4.4, the place of exports and imports of agricultural and food products in the Polish economy underwent some changes in the post-war years. In the early stages of rapid industrialization the share of agricultural and food exports in total exports was fairly high at around 17 per cent, while imports of agricultural products were a relatively small share of total imports at 11 per cent.

TABLE 4.4 Share of agricultural and food products in Poland's foreign trade, 1950–74

| *Years* | *Exports* (*percentage*) | *Imports* |
|---|---|---|
| 1950–6 | 17.3 | 11.1 |
| 1957–65 | 17.8 | 14.7 |
| 1966–9 | 16.1 | 13.4 |
| 1970–2 | 14.0 | 12.8 |
| 1973 | 15.0 | 13.0 |
| 1974 | 12.6 | 12.4 |

In the years 1957–65 the Polish economy was opened up more to foreign markets. As regards agricultural and food products, their share of imports rose from 11.1 to 14.7 per cent and the share of exports rose slightly. The ratio of agricultural exports to imports fell to 1.09. After 1965 there was a rapid decline of the share of agricultural and food products in both exports and imports of Poland. The volume of trade in these products has grown considerably (between 1960 and 1974 the volume of exports and imports of agricultural and food industry products increased by 3.3 and 3.2 times respectively). Their share in total exports and imports declined, however, following a rapid growth of industrial exports. The relation of exports to imports in the years 1966–9 rose slightly to 1:1.15 and in 1974 was near to parity.

In the years 1965–70 persistent attempts were made in Poland to improve the proportions between exports and imports, especially by savings on imports of agricultural commodities, primarily fodder. Consequently, the rising share of imports of agricultural goods and production means for agriculture in the country's total imports was checked. However, checking the growth of fodder imports brought about many difficulties in maintaining the upward trend in the number of pigs, which in turn considerably complicated the food supply in the country. Thus, apparent savings in fodder imports caused serious losses in agricultural production.

After 1970 Poland abandoned that policy and increased fodder imports. This allowed agriculture to use considerable reserves of potential production, which would have never been revealed without fodder supply from outside. These fodder imports enabled exports of animal products to be increased. Exports of animal products from Poland are in part re-exports, as they are based, among other things, on imported fodder. However costly these imports have been, particularly in recent years, they are beneficial because they allow agricultural output to be increased and a fuller use to be made of resources and reserves available in agriculture (manpower, buildings, output capacity of meat industry, etc.).

The policy implemented in Poland after 1970 resulted for the first time in the post-war period in the value of imports exceeding the value of exports. This was the result of a wider opening up of the Polish economy to foreign markets and of the changes in development strategy, in which imports were recognized as an independent factor accelerating the country's economic advance.

Grains and high-protein feeding stuffs are major imports, constituting over 40 per cent of total agricultural imports. An increase of grain imports by 30 per cent in the years 1970–4 and the doubling of the imports of high-protein feeding stuffs were linked with the needs of the dynamic growth of animal production, which increased in those years by about 30 per cent. Poland also exports minor quantities of grain. Taking into account the net imports of grains, the degree of Poland's self-sufficiency in grains was 87 per cent in 1974.

The major item in Polish agricultural and food exports are products of animal origin. For many years Poland has held the position of an outstanding exporter of these products to the world market. Animal products account for some 60 per cent of the value of total exports of agricultural and food commodities. The favourable trade balance in animal products covers the deficit of the trade in crop products.

Polish agriculture is actively adjusting to the changing situation in world markets as is shown by the following major examples. According to changes in the demand structure for meat and meat products, operations were undertaken in Poland aimed at increasing production of the high-quality beef in demand. High-quality pedigree cattle are imported, wide-scale breeding operations are carried on, a careful selection of cattle made according to more demanding quality criteria, feeding is being improved and the farm industrial fattening of young slaughter cattle is being introduced on a large scale. These are large-scale operations which it is expected will bring long-term results.

Another example of the adjustment operation is the sugar policy launched recently. The increasing demand for sugar and the consequent rise in world prices brought about in Poland a sequence of decisions of long-range significance. The aim is to increase the sugar beet cultivation area (the purchasing prices for sugar beets were raised), to improve cultivation and seed management, to improve raw material transportation and to build new, and modernize existing, sugar factories, etc. The aim of these operations is to produce 2200,000 tons of sugar yearly in 1980, the major part of which will go for export.

Cooperation arrangements amongst CMEA countries provide for a degree of specialization of production and trade. Poland, for instance, is a major provider to other CMEA countries of high-quality potatoes and other seeds, and imports part of its grain requirements from the USSR. The production of tractors and some other farm machinery is specialized by type among member countries, as are also some lines of agricultural research. In this way, agricultural adjustment in the CMEA countries can benefit through intra-trade from economies of scale in the output of farm machinery and from advantageous location of production of certain commodities.

In the difficult world food situation, and in view of the growing food problem in a number of food deficit countries, Poland assists by considerably increasing her agricultural production, a significant part of which is designated for export. In addition, Poland is an exporter of a certain amount of fertilizers (mainly nitrogen fertilizers), some agricultural machines, grain combines and tractors, and in this way she contributes to the development of the world agriculture.

NOTES

1 A 'voivodship' is an administrative unit; until June 1975 Poland was divided into 17 voivodships, and subsequently into 49.

2 Chapter 7 examines in detail the implications of these future changes.

# 5 Socio-economic Structure of Agriculture: Evolution and the Present State

## GENERAL CHARACTERISTICS OF THE SOCIO-ECONOMIC STRUCTURE OF AGRICULTURE

In the post-war period in Poland the socio-economic structure of agriculture was strongly influenced by progress in the industrialization of the country and in the socialization of agriculture. The expanding industrial sector exercised a far-reaching impact both on agricultural production and on structural changes in agriculture. Before the war there had been a marked trend towards fragmentation of individual farms as a result of the partitioning within families and of the absence of an outflow of rural population to non-agricultural jobs. After the war there was also an increase in the number of individual farms, but the mechanism was different. Whereas before the war fragmentation was a consequence of land hunger and increasing agrarian over-population, after the war it resulted from the rapid industrialization of the country.

There were two sources of the growing number of smallholdings. Some farmers' families disposed of a part of their land in order to fit the farm area to the amount of manpower available. For others, the piece of land became a home plot, and the main income was provided by work outside agriculture. In this way there was formed a type of farm of an auxiliary, food-providing nature involving the manpower surplus of the farmers' families. This is the group of rural dual-occupation holders, referred to in Chapter 4.

In Polish agriculture there are three forms of ownership of the means of production, namely individual, cooperative and State. They result in three types of farms. The ownership does not, however, always coincide with usage of the land because a part of the State-owned land (about 1m hectares) is used by agricultural cooperatives, agricultural circles, individual farms, as well as by the non-agricultural population.[1] State-owned land, transferred free of charge for use by agricultural cooperatives, becomes cooperative property, but if the cooperative should later be dissolved the land is taken over again by the State.

The distribution of agricultural land among different users in 1974 was as follows:

| | *Area* (*in '000 ha.*) | *Percentage* |
|---|---|---|
| Individual farms | 15,450 | 80.2 |
| State farms | 3203 | 16.6 |
| Cooperative farms | 288 | 1.5 |
| Farms of agricultural circles | 155 | 0.8 |
| Others[a] | 161 | 0.9 |
| Total: | 19,257 | 100.0 |

[a] This is agricultural land of individual users of plots, under 0.1 ha., and State land not included in farms as, for example, land in forest complexes used for fodder production for forest animals or used by the employees of these complexes and for home plots.

In spite of far-reaching changes in the structure of Polish agriculture over the past thirty years, individual farming still predominates. In 1974 there were 3,208,000 individual farms of over 0.5 ha. operating in Poland. They occupied 80 per cent of the land area and produced 81.5 per cent of the total value of agricultural production, including 77 per cent of the marketed production. The effect and economic significance of the individual sector is, however, diminishing, particularly in recent years. Individual farms use about 600,000 hectares of State land leased from the so-called State Land Fund. This fund is made up mainly of land transferred by elderly farmers without successors to the State in return for a pension, or of land taken over by the State after the death of the owner when there are no successors entitled to inherit. (Children who are not farmers do not have the right to inherit land in Poland.) This land is transferred for use to the State and cooperative farms as well as to agricultural circles, or it is leased, and sometimes also sold, to individual farmers.

The socialized sector in Polish agriculture is represented by 7178 State farms and 1061 agricultural cooperatives, covering in all 20 per cent of the agricultural land and providing just under 23 per cent of the marketed production. Its share in the total production of agriculture is somewhat smaller at 18.5 per cent.

The size and output of State farms and of agricultural cooperatives do not, however, give a full picture of the place of the socialist sector in Polish agriculture. There are in addition 35,248 agricultural circles, operating in 88 per cent of the villages, and 2249 inter-circle machine bases, which employ 168,500 persons and have at their disposal large social property, i.e. 35 per cent of the total number of tractors and a considerable part of the agricultural machines. The production assets of agricultural circles are social property and serve the intensification of agricultural production of all farms in single villages. These assets now exceed 39 billion zlotys and constitute an essential element in the social ownership of production means in agriculture. On the basis of this production potential and of experts employed by agricultural circles, large-scale instruction activities have been developed. Moreover, agricultural circles engage in group production operations, using 155,000 hectares of land for this purpose.

Besides State farms, agricultural cooperatives, agricultural circles and inter-circle machine bases, essential elements of the socialist sector in agriculture include: State machine centres and their agencies, numerous agricultural unions, trade associations, State enterprises for electrification of agriculture, veterinary and animal insemination stations, and a number of other enterprises rendering production services to agriculture. All of these have a considerable social capital at their disposal.

There are also rural marketing cooperatives which, through commercial channels, integrate individual farming with the national economy and socialize the marketing of agricultural products. They supply agriculture with production means as well as providing the rural population with essential consumption goods. The marketing cooperatives, together with agricultural food processing enterprises, form effective channels for vertical integration of individual farming with all the other sectors of the national economy.

Thus, a discussion of the place and role of the socialist sector in Polish agriculture cannot be limited to State farms and agricultural cooperatives only. If not only land but also other production means (such as agricultural machines and equipment, building, transportation, etc.) which play an ever-growing role in modern agriculture are taken into account, it can be seen that the socialist sector in Polish agriculture is relatively strong, and certainly much stronger than would appear from the share of socialized land, which is still not very large.

## REGIONAL DIFFERENTIATION OF AGRARIAN STRUCTURE

The agrarian structure in Poland shows a considerable differentiation amongst regions. From this viewpoint, the following four regions have been singled out by the Institute of Agricultural Economics:

Region 1 South-east region (voivodships of Katowice, Cracow, Kielce, Lublin and Rzeszów);
Region 2 Central-east region (Lódź, Warsaw, Bialystok);
Region 3 Central-west region (Gdańsk, Bydgoszcz, Poznań, Wroclaw, Opole);
Region 4 North-west region (Olsztyn, Koszalin, Szczecin, Zielona Góra)[2]

These regions differ from one another with respect to both the share of socialized (State and cooperative) farms in the total area of agricultural land, and to the average area of individual farms.

The future development of agriculture will not be the same in, for instance, Regions 1 and 4 respectively. Region 1 has a highly dispersed agrarian structure; individual farms are typically small, and large farms (State and cooperative) cover only 6.2 per cent of agricultural land. In Region 4 by contrast, nearly half of the agricultural land is in large farms, and the average area of individual farms amounts to about 10 hectares.

It is characteristic that changes in the agrarian structure proceed much more slowly in Regions 1 and 2, with a dispersed agrarian structure, than in Regions 3 and 4, where large farms prevail. In these latter regions, there occurs a faster outflow of people from agriculture and of farms from production. This leads to a polarization of farms into two groups, with an increasing area occupied by larger and more efficient farms, and a declining area occupied by part-time (or 'dual-occupation') farms.

TABLE 5.1 Regional differentiation of agrarian structure, 1973

| *Item* | *Regions* | | | | *Total* |
|---|---|---|---|---|---|
| | 1 | 2 | 3 | 4 | |
| | | *(percentage)* | | | |
| Share of socialized farms in the use of land | 6.2 | 6.0 | 27.4 | 47.7 | 19.5 |
| | | *(hectares per farm)* | | | |
| Average area of an individual farm in hectares of agricultural land: | | | | | |
| over 0.5 | 3.4 | 5.7 | 5.7 | 7.3 | 4.8 |
| over 2 | 4.6 | 6.8 | 7.8 | 9.8 | 6.3 |

The different situations in the various regions call for appropriate adaptations of agricultural policy to stimulate the desired changes in agriculture. In regions with a dispersed structure, agricultural policy aims at polarization of farms and the improvement of agrarian structure inside individual farming. On the other hand, in regions with a better agrarian structure the target is to slow down the number of farms going out of production by, among other things, improved means of investment and working credits. However, since farms do go out of production, the task of agricultural policy is to assure the most effective use of the vacated land. In these regions there is promoted, in particular, the development of the socialist sector in agriculture as well as different forms of common farming by groups of individual farmers.

INDIVIDUAL FARMS

At present there are in Poland over 3.2 million individual farms, excluding small plots of less than 0.5 hectares, with an average of about 4.8 hectares of agricultural land. Most of the farms from 0.5 to 2 hectares are auxiliary or part-time farms of dual-occupation holders, except for a small number of intensive vegetable farms. Farms of up to 2 hectares are not considered as regular agricultural workshops. Disregarding these, the average area of a farm amounts to about 6.3 hectares of agricultural land. The data on structure of individual farms in 1974 are shown in Table 5.6.

Ten hectares is considered to be the area that, under the present conditions in Poland, ensures both full use of family labour resources and an adequate family income. Some smaller-sized farms intensively operated in specialized production of vegetables, fruit, tobacco, poultry or pigs, also produce high incomes. In recent years the agrarian policy has encouraged farm specialization through investment credits on easy terms. (This subject is discussed later on pp. – .) At present, occupants of not more than one million farms in Poland earn their incomes from agricultural production only. Occupants of nearly two million farms have two income sources – from agricultural production and from paid work elsewhere. One million of these are part-time farms, i.e. the head of the family is employed outside agriculture, and the other million are farms in which one family member is employed outside agriculture. The number of persons employed on farms varies considerably according to the size of farm, ranging from 0.31 persons in the group of 0.1 – 0.5 ha. to 2.80 persons in the group of over 20 ha. The range in the number of men is still wider, from 0.04 to 1.45 respectively.

TABLE 5.2 Structure of individual farms, 1974

| *Area size* | *Number of farms* | | *Area of agricultural land* | |
|---|---|---|---|---|
| *(hectares)* | *('000)* | *(%)* | *('000 ha.)* | *(%)* |
| under 0.5 | — | — | 93 | 0.6 |
| 0.5–2 | 910 | 28.4 | 1004 | 6.5 |
| 2–5 | 995 | 31.0 | 3183 | 20.6 |
| 5–7 | 442 | 13.8 | 2487 | 16.1 |
| 7–10 | 437 | 13.6 | 3399 | 22.0 |
| 10 and over | 424 | 13.2 | 5284 | 34.2 |
| Total | 3208 | 100.0 | 15,450 | 100.0 |

In 69 out of 100 farms in the 0.1 – 0.5 hectare group, there are no persons working on the farm only, while in 97 there are no men working on the farm only. The number of farms without men permanently working on them declines gradually with the increase of the farm area. However, even in the 5–7 hectare group there are no men permanently working on 7 per cent of the farms, and there are no men of productive age working on 10 per cent of the farms. This reflects the dual-occupation nature of individual smallholdings and the much higher proportion of men than of women engaged in employment outside agriculture. According to the statistical data, the dual-occupation phenomenon disappears only in farms of over 10 hectares. This proves that an individual farm with too limited production potential to catch up with and to employ more technically advanced methods is unable, even with a fairly high price level for agricultural products, to achieve an income level comparable with that of persons employed outside farming. The reason is primarily low labour productivity.

In farms of under 10 hectares the number of professionally active women working on the farm exceeds the number of men, showing that a large number of farms are worked mainly by women. Disregarding plots of up to 0.5 hectares, 37 per cent of the farms are on the average worked by women. The share of farms worked by women, classified according to the size of farm, is as follows:

| | |
|---|---|
| 0.5– 2 hectares | 80 per cent |
| 2– 5 hectares | 45 per cent |
| 5–10 hectares | 17 per cent |

In individual farms over 10 hectares and in cooperative farms the ratio between the number of working women and men is one to one. In State farms the number of women employed is only half that of the men employed, because many work only on their own home plots.

The important role of dual occupation among individual farmers in Poland is indicated by the fact that in the years 1960–70 the number of persons employed outside their farm increased by one million, from 1.7 to 2.7 million. The share of income from paid work outside the farm in the total income of agricultural families is also growing, having increased from 19 per cent in 1960 to 31 per cent in 1970. This share in 1970 was as follows, classified according to the size of farm:

| | |
|---|---|
| 0.5–2 hectares | 63 per cent |
| 2–7 hectares | 33 per cent |
| over 7 hectares | 11 per cent |

The development of a dual-occupation system for the agricultural population has so far been beneficial because work outside agrcilture has used surplus manpower and has provided additional income for the farm population. However, there are disadvantages from dual occupation, in particular slower production growth on farms of dual-occupation holders as compared with purely agricultural farms. They constitute a barrier to the technical modernization of agriculture and limit the possibilities for growth of social labour productivity. Economic policy therefore favours the polarization of dual-occupation farms towards either reducing the farm size and leaving just a small home plot, or towards expanding the farm size until it is adequate for a purely agricultural farm.

## STATE FARMS

In 1974 State farms covered 3,203,000 hectares of agricultural land. Since 1960 this area has grown by 800,000 hectares and now increases by about 70,000 hectares yearly. It is anticipated that in the next five-year period the area of State farms will grow by about 120,000 hectares yearly. State farms are taking over land vacated by individual farmers when it is situated in the vicinity of existing ones. New farms are being set up when at least 200 hectares of vacated land are available in one complex.

Out of the 3,203,000 hectares owned by State farms, 2,973,000 hectares are used by farms under the Ministry of Agriculture and the remaining 230,000 hectares are operated as experimental farms for agricultural schools and universities and other institutions.

State farms of the Ministry of Agriculture are under the management of the Director-General in the Ministry of Agriculture. He exercises control over sixteen regional boards, four national boards associating specialized farms (plant breeding and seed production, cattle breeding, horticultural seed production and nursery gardening, fisheries), regional experimental farms (one or two in each voivodship) and farms run by agricultural institutes of the Ministry of Agriculture.

There are at present 630 agricultural State enterprises controlled by Voivodship Boards, with an average area of 2673 hectares of agricultural land and the size ranging from 300 to over 7000 hectares. On average, one enterprise includes 4.2 farms (production units). However, there are both one-unit enterprises and enterprises covering ten units each. One-unit enterprises are found mainly in regions with few State farms. The merging of farms into larger agricultural enterprises aims at providing better conditions for the specialization of single farms within the framework of the enterprise, and at increasing the scale of production.

State farms have experienced in their history many changes in the management system. In the fifties they were operated under a system of centralized management. The operation plans were prepared by single farms but were approved and often revised by superior authorities. Under the present system, State farms enjoy considerable freedom in shaping their activities; they prepare economic plans but for their own use only. The director of a State farm is now bound only by two directives coming from superior authorities, namely the wage fund and the value of net commodity production,[3] but it is the director of the enterprise who decides what and how much will be produced. The wage fund can be exceeded in case of above-plan growth of net commodity output. According to the market situation existing in the country or in the given region, central or regional authorities recommend production increases for specified articles. These are not directives, however, and are accepted only if they do not interfere with regular operation of the enterprise and its production pattern.

The role of State farms in Polish agriculture is concentrated in the following three fields of activity:

provision of agricultural production means for agriculture as a whole;
participation in the changing of the agrarian system;
agricultural commodity production.

In the existing agrarian structure and State responsibility for development of agricultural production, the first area of activity imposes on State farms obligations which are not included in the regular operation of enterprises that seek to maximize their profits. High-quality seeds and animals for breeding purposes produced by State farms are sold to farmers at prices lower than their actual production costs and thus subsidized by the State for the purpose of raising farming standards and productivity generally. State farms are an almost exclusive producer of quality seeds of cereals and of many other plants, as well as of animals for breeding purposes. For example, of the total 1973 production of 615,000 tons of quality seeds of cereals, 584,000 tons or 95 per cent were provided by State farms.

The State farms also participate in changing the agrarian system. In the present agrarian structure of the country, State farms are the main representative of socialist large-scale farming. Increase of the area under cultivation by State farms is the chief form of growth of the socialist sector in agricultural production. The development of the socialist sector is, in this respect, of a rather passive nature; it is the result of processes occurring inside individual farming and leading to land dropping out of use. What is socialized in this event is not the farm, but land without farmers, production means or manpower. Recovery of the land for agricultural production requires additional costs for restoring the culture of the soil, production equipment and management of the area. Thus it is a costly undertaking for the national economy, but it is indispensable not only for the reconstruction of the agrarian system but for continuation of the necessary growth of agricultural production.

In the case of State farms, the increase of their area by the absorption of individual farms has not yet been practised but it is not out of the question for the future. The developing forms of productive cooperation between individual farms and State farms can pave the way in the future for individual farms to join the enterprises with which they have cooperated for some time to their mutual benefit. On the other hand, regardless of future decisions of individual farmers, the cooperation provides a form for expanding the production opportunities of both parties, the production of State farms grows without an otherwise necessary expansion of their area and without costly investments in real estate, and individual farms have the opportunity for long-term specialization on the basis of requirements of the big neighbour.

The State farms also have a further role in the reconstruction of the agricultural system – namely, searching and experimenting in forms of socialist organization of agriculture in the macro-economic sense, i.e. enterprise, services, settlement and social welfare infrastructure, administration, and the like.

The third field of activity for State farms is agricultural commodity production. Their share in agricultural commodity production continues to rise, from 12.7 per cent in 1960 to 18.2 per cent in 1973. Their share in commodity production is now higher than that in the area of agricultural land, which amounted to 16.4 per cent in 1973. A very significant role is played by State farms in the production of cereals. In 1973 State farms supplied 1.75 million tons, i.e. 30 per cent of the 5.89 million tons of cereals purchased. State farms are also the main supplier of young slaughter cattle for export.

## ARICULTURAL COOPERATIVES

In 1974 there were 1061 cooperative farms in Poland with an area of 288,000 hectares of agricultural land. Of this total, 276,000 hectares were used in common farming, while 12,000 hectares were used individually by the cooperative members. Every family in the cooperative is entitled to own a home plot of up to 0.5 hectares and to keep livestock, namely pigs, poultry, sheep and cattle. Depending on the decision of the general meeting of members, the number of cattle is usually limited to two cows with their young. In the majority of cooperatives which achieve higher incomes through common farming, the members voluntarily give up the keeping of their own cows as it becomes unprofitable and too burdensome. In 1960 cooperatives had on the average 1.1 cows per family, but by 1973 the figure had fallen to 0.5. The proportion of the animals kept by the members individually shows a regular decline. While in 1960 63 per cent of the entire livestock of the cooperatives was kept by the members individually, by 1973 the share was only 16.7 per cent.

As shown in Table 5.3, the number of cooperatives decreased from 1668 in 1960 to 1064 in 1973. This was not tantamount to a dissolving of cooperatives, although such cases did occur, but rather to the merging of small cooperatives and the setting up of larger farms. The total area of cooperatives in the country grew during that period by 36 per cent, while the average area per cooperative rose by 112 per cent, and the average

TABLE 5.3 The development of agricultural cooperatives

| *Specification* | *1960* | *1970* | *1973* |
|---|---|---|---|
| Number of cooperatives | 1668 | 1071 | 1064 |
| Total area of agricultural land of cooperatives (thousand ha.) | 205 | 246 | 278 |
| Average area of a cooperative (ha. of agricultural land) | 123 | 230 | 261 |
| Total number of families in cooperatives (thousands) | 22.4 | 24.8 | 31.2 |
| Average number of families per cooperative | 13.4 | 23.2 | 29.3 |
| Total number of animals per 100 ha. of agricultural land | | | |
| cattle | 40.3 | 54.4 | 77.9 |
| cows only | 25.5 | 20.8 | 20.2 |
| pigs | 62.5 | 48.0 | 95.6 |
| Number of cows kept individually per family | 1.1 | 0.7 | 0.5 |
| Percentage of total livestock in cooperatives kept by members individually | 63.0 | 24.1 | 16.7 |
| Cereal yields (quintals per ha.) | 18.0 | 22.2 | 30.4 |
| Total personal income per family (thousand zlotys) | 32.5 | 49.7 | 72.3 |
| Personal income per family from group farm (thousand zlotys) | 17.7 | 36.0 | 55.6 |

number of families per cooperative by 120 per cent.

The data show a dynamic growth of livestock per hectare and of cereal yields, which was reflected in the rise of agricultural production and of incomes per family. In the years under study, these incomes increased in current prices by 122 per cent, or by 108 per cent in constant prices; the entire increase was in income from the group farms, since income from the individual farms of cooperative members showed no increase.

In 1973 there were about 9 hectares of agricultural land per family associated in cooperatives. However, the land area per family varied considerably amongst regions of the country and ranged from 5 to 15 hectares. Where the area per family is too small to provide sufficient income from the agricultural production of cooperatives, there are various kinds of non-agricultural activities. These may include vegetable and fruit processing enterprises for which cooperatives may also buy under contract produce from individual farmers, building material production enterprises, and many kinds of services – for example mechanization and transportation services, and even small-scale industrial products. On the average about one-third of the cooperative incomes are derived from non-agricultural activity and in small-size cooperatives the share may be as high as 60 per cent per family.

However, the development of non-agricultural activities in cooperatives cannot provide a solution for the future for the growing number of cooperative farms. As a result of their use of a higher proportion of purchased inputs for agricultural production, cooperative farms achieve lower income per area than do individual farms. Moreover, a much bigger part of this income is allocated for fixed capital formation and indivisible funds.[4] In order to ensure the present income level per cooperative family from agricultural production alone, the average area per family should amount to at least 12 hectares of agricultural land.

The advance of cooperative farming in Polish agriculture is still influenced by the unfavourable experiences of the fifties when, in the organization of cooperatives, the principle of voluntary membership had not always been observed. This situation, however, is beginning to change slowly. After several years of stagnation, 6500 families joined existing or newly organized cooperatives during the period 1971–4, indicating that some reluctance on the part of farmers to agricultural cooperatives is being overcome.

Until 1972 two types of cooperatives operated in Poland, namely the higher type with fully common crop and animal production (except for home plots), and the lower type, with partly common production, usually crop production, and with animal production tending to remain an individual activity. This, of course, did not rule out a gradual development of a common livestock activity.

For quite some time cooperatives of this lower type were underestimated in Poland. These cooperatives, considered as not being fully socialist, were induced to go over to the higher type, which in most cases did not help their development. Practice indicated that this form of cooperative farm is much more acceptable for individual farmers, as it does not radically change their status of 'farmer'. At the same time, it allows the combination, particularly at the initial stage, of the advantages of group and individual farming. It is also a form which, with the experience of advantages of group farming and the inconveniences and limitations of individual farming, will almost automatically undergo further transformation.[5]

Changes introduced in 1973 in the firmly applied statutory rules, for example the reduction of the minimum number of families from the previous ten, make possible now the organization of cooperative farms by even five families, sometimes by even four families, and give considerable freedom in the choice of directions and scope of group and individual activity. Indeed, the new simple forms of agricultural cooperatives differ little, apart from the formal legal aspect, from group forms of farming as discussed below.

## GROUP FARMING

New opportunities and a new path of development for individual farming in Poland was provided by the law of 11 February 1972 on cooperation in agriculture. This law set up legal bases and ensured material means for new organizational forms in individual farming, ranging from the simple forms of cooperation to full integration of individual farms.

Simple forms of productive cooperation among farms include for example the common use of machines and buildings and cooperation in animal or crop production, common cultivation of pastures, cultivation blocs for some crops, common cultivation of land of the State Land Fund, and common breeding of young slaughter cattle or fattening of pigs. A characteristic feature of all cooperation forms is that only certain functions of the farm are subjected to the common interests of the whole group, without encroaching on the actual status of the individual farms.

The simple forms of productive integration among farms are production undertakings run in common and as a joint venture by a group of farmers. The integration of farms can be partial, i.e. it can involve one or several production sectors, or it can be full, consisting of total merging of farms and the setting up of new, common farms. The full integration of farms is, of course, the highest form of those mentioned. It means that as regards production decisions independence is substituted by common, collective decisions. It allows the introduction of division of work and the benefits from this division.

The reasons for creating new organizational forms of individual farms were both economic – technical and social. Considerable fragmentation of individual farms does not allow many of them to be equipped with technical installations and implements which are indispensable for the mechanization of agriculture. Only the bigger farms of about 10 hectares and more can afford to purchase tractors and some machinery. On the other hand, farmers holding smaller farms can purchase such equipment only if they form a group of several persons whose aggregate ownership of land would allow full use of this equipment.

Organizing farmers groups facilitates modernization of production, mainly through specialization along determined lines. Economic benefits which can be achieved by farmers organized in groups are particularly the following:

- increase of production and of labour productivity, as well as the raising of incomes obtained from the farm;
- the opportunity for adequate technical equipment of farms with a simultaneous lowering of mechanization costs;
- the possibility of increasing the scale of production by taking over State land and buildings, and also those of farmers who are unable to use them fully.

Farmers organized in groups can take advantage of assistance available from the State in the form of credits,[6] supply of building materials and technical equipment, as well as extension services.

Such a form of small groups (the law provides for a minimum of three farmers) is much more acceptable to farmers than the present form of agricultural cooperative. In less than three years after the law came into force, about 6300 production groups were established, associating about 21000 farmers (data as of 31 October 1974). This form has therefore been accepted by Polish farmers and promises to develop further.

Groups of individual farmers have priority in taking on lease land from the State Land Fund and thereby expanding the acreage of their common farm. As a result of taking over additional land, the average acreage per family associated in groups increased from 7.3 to 9.3 hectares of agricultural land. The average acreage per group amounts at present to about 28 hectares.

The short period of activity of the groups does not yet allow a full evaluation of their production and economic results. Most of the groups are oriented towards specialization in one of the selected production lines. The extent of integration of the groups varies. In some the integration is only partial, covering one production sector (for example, common production of cattle, pigs, sheep or poultry). In others it covers several lines, including common cultivation of land taken on lease, and in some groups the integration is full, which means common running of both crop and animal production.

The provision of various forms of gradual changes in individual farming (beginning with simple forms of cooperation, through group forms, and winding up with advanced forms of agricultural cooperatives) gives to farmers willing to exceed the limited framework of individual farming a range of choice as to the form which is at the time the most convenient and acceptable to them. The fact that small groups can be organized provides opportunities for selecting partners who trust one another and of setting up groups which ensure effective collaboration. With the development and consolidation of small groups, conditions will arise favouring their integration within larger organizational units.

## FARMS OF AGRICULTURAL CIRCLES

Farms of agricultural circles are set up on land vacated by individual farmers, mainly in cases when there are no other opportunities of cultivation of this land. In 1974 the farms of agricultural circles covered 155,000 hectares of agriculturally productive land. It is anticipated that in the coming years this farming will be considerably expanded and will become one of the forms of socialization of agriculture in Poland.

Until now these farms were run by hired workers, engaged by the administrative board of the agricultural circle. In 1974 the legal status of these farms was however changed into group farms of agricultural circles, a form similar to cooperatives. They are now open farms, which means that they can be joined by individual farms. They differ, however, from the cooperative form by the fact that they are not independent legal units but are subjected to the community cooperative of agricultural circles.

As noted earlier, the cooperative of agricultural circles is a new organization, having been gradually introduced since 1973. It is an enterprise of a cooperative type, which covers by its operations the area of one community. The members of this enterprise are rural organizations of agricultural circles, and its operations consist of rendering mechanization and other services to farmers, repairing agricultural equipment, providing construction services, and organizing farmers.

It is assumed that the cooperatives of agricultural circles will gradually expand their operations and will become coordinators and then integrators of agricultural production in the area covered by their activity. They have a chance of changing in the future into community agricultural enterprises and of playing an essential role in transforming individual farming in Polish rural districts, as discussed in the following chapter.

NOTES

1 The non-agricultural population use the so-called home plots. These are areas allotted in towns or their vicinity, divided into plots of 300–400 square metres. The total area of these home plots in the country amounts to 19,500 hectares, and the number of plots to 427,000. These plots serve both recreation and production purposes; they are used by the urban population for vegetable, fruit and flower production.

2 According to the administrative divisions applicable after 31 May 1975.

3 This is the value of production sold, less the value of purchased production means of agricultural origin (seeds, fodder, animals for breeding purposes).

4 For example, in 1973 the value of savings and indivisible funds in agricultural cooperatives amounted to 2200 zlotys per hectare while the value of savings on individual farms amounted only to 1150 zlotys.

5 In 1960 there were 719 cooperatives of the lower type, while in 1973 the number had fallen to 126.

6 Group farms are granted investment credits on the following terms: period of payment up to 20 years, suspension period up to 4 years, interest 1 per cent on credit for cowsheds and 2 per cent for other credits. Moreover, credits up to 40 per cent of the costs borne for farm buildings are authorized to be written off.

# 6 Planning

## PLANNING IN AGRICULTURE AND MANAGEMENT OF ITS DEVELOPMENT

Management of an agriculture of such a complex structure as now exists in Poland is a very difficult task. There are 8800 socialized farms and 3.4 million individual farms of very differentiated socio-economic and professional structures.

The operation side-by-side of this variety of enterprises gives rise to a complicated structure of mechanisms for managing and controlling production. State farms are managed directly within the framework of a central plan. For the cooperative and individual farms, methods of indirect influence are adequate. Although the State influences individual farming through a number of economic instruments, it cannot control its development directly. This does not mean that production of the individual sector is not a subject for central planning but it is so only for government authorities and for enterprises servicing agriculture. It is not an obligatory plan for single farmers.

The directive part of the development plan for agriculture concerns the supplying of agriculture with production means of industrial origin and the carrying out of basic investments in agriculture. It also concerns the charges on agriculture for the benefit of the State, in connection with which there are planned directives as to the tasks of the economic apparatus in the purchasing of agricultural products and the processing of them by agricultural food industry.

There are three essential instruments by way of which the State influences individual farming, namely: prices of agricultural products and of production means for agriculture, together with interest and land rent; contract-buying; and investments.

The workings of the plan reach the agricultural producers fastest and most directly through prices, i.e. change of their respective levels and the relations between them. Both the purchasing prices and prices for production means bought by farmers are fixed by the State and are legally valid. These prices reflect the preferences of the State as regards production trends, dynamics and structure. Prices in this system of management have the character of parameters and are instruments whereby the State influences the economic motivation of the agricultural producers.

Poland follows a policy of stable prices. It does not mean, of course, that prices never change. The State changes purchasing prices and prices for production means rather seldom, however, and then only if conditions for production and market balances have changed essentially or the price change is needed to implement socially vital aims. Price stability is of fundamental significance for farmers; if they know the general price level and relations between prices, they can plan not only their production but also their investments and can allocate their resources rationally. Consequently, price stability means stability of economic conditions for production and of the material situation of the farmers.

Price policies are the State's instruments for implementing two main aims with respect to agriculture, namely those concerning production and the allocation of resources, and those concerning the incomes of the agricultural population. In their first function prices are a fairly effective instrument, although not the only one for shaping the level and structure of agricultural production and of the allocation of resources (including investment outlays in individual farming). In the other function prices are almost the exclusive instrument for shaping the incomes of the agricultural population. Decisions concerning production and allocation of resources involve mainly relations between prices for individual agricultural products. Those concerning incomes involve changes in the absolute prices, particularly as these affect the relationship between prices for agricultural products and those for production means.

In Polish agriculture, the role of credit is steadily increasing both as an instrument for controlling agricultural production and as the source for financing production and investments. This results from the concepts reflected in economic policy concerning the financing of net investment in agriculture, and from a favourable evolution of the economic situations of individual farms. With stable development and profitability of production assured, farmers are increasingly inclined to use credit as a source of means to satisfy not only current but also investment needs. Credits are granted for production investments, particularly for farm buildings and building repairs, for purchase of production working assets (fertilizers, production services), purchase of agricultural machines and equipment, for housing construction, etc. Over 80 per cent of the credits are designated for financing the development of agricultural production. Low interest rates on credits (1 to 6 per cent) are obligatory. The interest rate is usually lower than or equal to that paid on savings deposited in a bank. About 15 per cent of the credits are not subject to interest. Farms producing within the contract-buying system enjoy preferential treatment in obtaining credits.

However, price instruments and credit, though most effective, are not sufficient by themselves for the management of agriculture. As shown by economic practice, it is not enough to raise the purchasing prices in order to increase production. Agriculture needs primarily an adequate stream of production means and services. Thus, decisive importance goes to non-price mechanisms such as investment policy, direct control of the stream of production means flowing to agriculture, organization of the production and marketing process, etc. In Poland the market is considered by its nature to be an imperfect mechanism. It does not contain sufficient criteria and parameters for economic decisions, either at the level of the national economy or at that of an agricultural producer. Therefore the price mechanism must be supplemented and strengthened by other, non-price instruments of agricultural policy. In the practical management and guidance of individual farming in Poland, a mixed system of co-existing indirect instruments (price) and direct ones, including primarily the contract-buying system and investments, has developed.

Contract-buying is a system by which millions of individual farms can be effectively included in the socialized, planned economy. Over 80 per cent of animal and plant commodity production is ordered and purchased by the State. The non-socialized market plays now a smaller and supplementary role. It is mostly of significance in the marketing of vegetables and fruit, dairy products, breeding livestock, horses, etc. A great expansion of contract-buying has taken place in Poland and it now covers nearly 100 agricultural and animal products. Through these channels the State is purchasing from farmers 70 per cent of their marketed production, including 45 per cent of the cereals, 52 per cent of the potatoes, 91 per cent of the pigs and 100 per cent of the industrial crops (sugar beet, tobacco, hops, endive, herbs, seeds, rapeseed, flax, hemp, etc.).

Contract-buying is a tested way of including farming in the implementation of economic plans. It ensures independence to the farmer in making economic decisions according to his economic interest, while including the individual farm in the system of the planned economy. It also links farming with the central plan as well as maintaining businesslike relationships on the basis of exchange profitability. In this way it facilitates association of the broad social interests with the current material interests of agricultural producers. Contract-buying is the main channel of vertical integration of agriculture with the national economy, particularly with the agricultural food industry. It allows maintenance of a necessary linkage between measures influencing agriculture with price policy, as well as combining economic non-price incentives (credit, agro-technical and zoo-technical advice, assistance with seeds, etc.) with the direct influencing of agricultural production. As it combines price policy with elements of direct organizational activity of specialized State and cooperative enterprises, it eliminates direct production control.

The contract-buying system at first covered industrial crops, agricultural products for export (needing observation of defined quality standards and of delivery terms) and crops requiring special stimulation in view of the social interest. Later, it also included seed cultivation and, finally, the products produced on the largest scale which in Poland are cereals and potatoes. Contract-buying in Poland is widespread both as to the variety and the volume of purchases. During a year there are now concluded in

Poland 13 to 14 million contracts with individual farms for production and delivery of various products. Since marketed production comes from about 3 million farms, the State concludes on the average four to five contracts yearly with each farm. Thus it is a big organization problem. The contracting enterprises, however, have available a specialized apparatus which has this complex problem under control.

Contract-buying agreements regulate in Poland, as elsewhere, the total relationship between the agricultural producer and customer. The contracts are concluded on a fully voluntary basis. The producer undertakes to produce and deliver at a fixed place, on a date agreed by the parties, a specified quantity of products of a determined quality. On the other hand, the contracting enterprise undertakes the following: to collect the contracted quantity of products; to pay the amount due according to the pre-established price; to grant credit to the producer for purchase of production means; to supply the producer with agreed production means (seeding or breeding materials, pesticides, certain specialized tools, building materials, etc.); and to grant to the producer agrotechnical and zootechnical help in the form of free advice and assistance.

The final recipient of the agricultural product is, in most cases, the agricultural food industry. Some of the industries (for example, sugar, tobacco, brewery, herbs, fruit and vegetable, poultry industries) have their own contract services and establish links directly with the agricultural producer. The interests of other industries and export enterprises are represented in relation to the farmers by universal trade cooperatives, which carry out purchases and contracting at the order of those industries. The agricultural trade cooperatives, which handle contract-buying of many products, do not always have available a sufficient staff of technical advisers. Their intervention in the production process is therefore much weaker than in the case of contracts concluded directly by the industry. However, when the production process is not too complicated, expert outside advice is not essential for the farmer.

Contract-buying performs a major role in the modernization of methods of agricultural production and in spreading technical advance in traditional individual farming. There is to be noted in Poland now a consolidation of the elements of production contracting, though many factors such as reduction in farm numbers or resistance of some farmers to technical innovations, etc. tend to slow progress in this field.

In spite of the many favourable features of contracting, it should not be overestimated. The contracting system does not play an independent role and it is not a panacea that would solve automatically the problem of balance between supply of and demand for agricultural products. Ensuring such a balance depends on over-all conditions of economic development and requires application of many measures influencing agriculture. If a rapid growth of demand for agricultural products, as is now being witnessed, calls for acceleration of the rate of growth in production and supply, favourable prices or an efficient contracting system will not be enough. In this case there must be investments in agriculture and in the sectors servicing it, intensive introduction of technical advance, increased outlays for infrastructure, etc.

Managing the investment process of a multi-sector agriculture, as in Poland, is very complex. Investment decisions must be made by all parties concerned, i.e. State, cooperatives and individual farmers. In Poland a co-partnership rule of the socialist State and farmers in the investment process has developed. The State participates primarily in investments needed from the broad social viewpoint, development and other essential investments, including outlays for the infrastructure. On the other hand, farmers implement such investments as are of direct interest to them, as well as small-scale investments. A special function of State investments is to mobilize and to bring into operation the farmers' own means. Thus, for example, 50 to 60 per cent of the investment outlays for improvements are covered by the State and the remainder by farmers who finance detailed investments, mostly through State credit. The investment help of the State can be effective only when it mobilizes the farmers' own efforts, not when it replaces them. Farmers' investment outlays often constitute a small but essential link in the chain of such State investment undertakings as electrification, road building and others, in obtaining the anticipated results.

The Polish view is that it would be useless and ineffective to leave all capital investment to individual farmers. Considering the essential significance of some kinds of investments, it is desirable to gather them into the hands of the State or of self-governing farmers' organizations, cooperatives, producers' companies, etc. In this respect a particular role is played by the Agricultural Development Fund, referred to earlier.

Of the total agricultural investment outlays in Poland, farmers now provide about 33 per cent, the Agricultural Development Fund 15 per cent and the State over 50 per cent. By taking over the major investment effort, the State gains the opportunity of influencing directly the development of individual farming, the shaping of trends in technical advance and the modernization of production processes.

Summarizing the above remarks on managing and guiding individual farming, it should be noted that although this component of agriculture cannot be the subject of direct planning implemented by State directives, it must nevertheless be indirectly managed by the State. The management concept implemented in Poland is based on the belief that by knowing the typical reaction of individual farms to market incentives, desired changes in production can be brought about within resort to administrative measures. This concept does not deny the significance of non-price measures of economic policy. On the contrary, it virtually assumes the need for other instruments of which the primary ones are the organization of the production process and of marketing as well as investment policy.

## FINANCING OF INVESTMENT AND COSTS OF TECHNICAL AND SOCIAL RECONSTRUCTION

The process of concentrating agricultural land and organizing larger production units, regardless of their social form, is very costly. Agriculture is unable by itself to cover the costs of structural adjustments or, according to terminology adopted in Poland, of the process of technical and social reconstruction of agriculture. This is because of labour productivity being lower in agriculture than in other sectors, the difficulties in ensuring income growth for the agricultural population equal to that for other professions, and the high material consumption and capital intensity of modern agricultural production.

The technical and social reconstruction of agriculture is in the interest of the whole society, since it releases from agriculture manpower needed for the growth of non-agricultural sectors. Manpower released from agriculture must be replaced by technical means of labour, and consequently substantial investment outlays are required. It would not be theoretically justified or socially equitable for the population remaining in agriculture to bear these outlays. Rather, they must be borne by the entire society and financed to a greater or lesser degree from State budgetary means. The State, by financing structural changes in agriculture, intervenes directly or indirectly in these processes, both for accelerating them and for orienting them according to the socio-political character of the socialist system.

The financing of structural adjustments in agriculture is implemented by direct or indirect subsidies. These subsidies cover direct financing of investments in the State sector, part of the investment outlays in cooperative and group farms, and tax reductions for investments in individual farms. For example, cooperative farms are subsidized in the form of the remission of credit for up to 60 per cent of construction costs of the first livestock building (cowshed or pigsty). Group farms obtain in such cases up to 50 per cent subsidy from the State budget. Cheap investment credit is also a form of subsidy. For example, the interest rate on investment credit granted to cooperative, group and individual farms amounts to 2 per cent, sometimes to 1 per cent. Thus it has not differed much from the inflation rate, which amounted to about 1.2 per cent yearly in Poland in the years 1960 to 1973. In addition to subsidies for investments State and cooperative farms which take over land for cultivation from the State Land Fund obtain subsidies of 15,000 zlotys for each hectare of agricultural land. These are subsidies for supplying the farms with working means indispensable for effective and intensive development of the newly acquired land.

Investments in the State sector of agriculture are financed from three sources: directly from the State budget, from the depreciation fund which is transferred by single State farms to a special bank account, and from a part of their own profits. Over 70 per cent of the investment outlays comes from the State budget, reflecting the increasing acreage of State farms (about 70,000 hectares yearly) and the need for additional investment on most of these farms.

In the sector of peasant farming (individual, group and cooperative farms) there exist two forms of investing, individual and social. The former are those investments made directly by farmers themselves. The social form covers three groups of investments, namely land improvement, electrification and, partly, mechanization. Land improvement and electrification are carried out by State enterprises and financed by the State budget. Farmers are charged with only a part of the costs of these investments which they pay off over ten years. Investment outlays for the purchase of machines in agricultural circles are financed from the Agricultural Development Fund, without taking account of the financial means of agricultural circles coming from allowances for depreciation and from profit.

The share of financial outlays made directly or indirectly by the State amounts to almost 60 per cent of total investment outlays for agricutlure, while the remaining part is covered from the resources of individual and cooperative farms. For example, in 1973 gross investment outlays for agriculture amounted to 62.9 billion zlotys (about 31 per cent of the value of gross national income produced in agriculture), of which 52.2 billion zlotys were for production investments and 10.7 billion zlotys for housing investments. Individual and cooperative farms themselves provided 21.3 billion zlotys, machine centres of agricultural circles provided 5.5 billion zlotys, and depreciation allowance from the Agricultural Development Fund and outlays financed by the State (including investments of State farms from their own resources) amounted to 36.1 billion zlotys.

State investments in agriculture are financed directly or indirectly from the State budget, with a minimum direct contribution of agriculture to this budget in the form of taxes. Total fiscal charges in 1973 met by peasant farms were the following, in billion zlotys:

| | |
|---|---|
| Land Tax | 5.2 |
| Community Tax | 1.4 |
| Agricultural Development Fund | 4.1 |
| Other charges | 1.2 |
| Total | 11.9 |

In comparison with the gross income from agricultural production of 167.9 billion zlotys, fiscal charges on peasant farms are very small, amounting to only 7.1 per cent. The absolute amount of these charges remains, in principle, unchanged, despite the rising income of peasant farms. For example, in the early sixties these charges amounted to about 13 to 14 per cent, in 1971 to 11.2 per cent, and in 1972 to 8.9 per cent of the income. Keeping the fiscal charges on peasant farms unchanged while incomes grow is intended to facilitate their own financing of ever-growing investment outlays. In 1965 the investment value on individual farms amounted to 740 zlotys per hectare, while by 1973 it had risen to 1410 zlotys. The annual rate of growth of investment outlays averages 9.6 per cent.

However, regardless of the income growth and the relative decline of financial charges, the peasant farms are unable to finance fully investments from their own means. A part of their investment outlays is financed indirectly by the State in the form of an adverse balance between the flow of new and of repaid credits. Thus in 1965, 23 per cent, and in 1973, 44 per cent of these investments represented increased indebtedness for the farms. This indebtedness, which grew in 1973 by a further 9.4 billion zlotys and amounts to about 60 billion zlotys, is the equivalent of about 26 per cent of the yearly value of agricultural production and about 36 per cent of yearly value of the gross income of the peasant farms.

Future growth of agricultural production, structural changes in agriculture and the reduction of employment in agriculture by about a half in the next fifteen years will require very large increases in investment outlays. In the years 1961–5 gross investment outlays in agriculture for production purposes (i.e. excluding housing and social construction) amounted to 98 billion zlotys in 1971 prices. In the years 1966–70 they increased to 161 billion zlotys. For the 1971–5 plan period, the figure will be about 250 billion zlotys.

Investment in agriculture has so far aimed chiefly at increasing production potential. In future, additional outlays will be allocated for the replacement of labour, which is leaving agriculture, by durable production equipment, and for reconstruction of the production means of individual farms no longer being operated. These outlays are required in order to integrate these farms into newly organized socialized and group farms or into individual farms which are enlarging their acreage. Both of these forms of investment outlays are the equivalent of investment costs of structural adjustments, while from the broad social viewpoint they are a part of the investment costs of attracting manpower from agriculture to the non-agricultural sectors of the national economy. According to estimates of the Institute of Agricultural Economics, the necessary investment outlays to replace one full-time worker leaving agriculture amount to about 300,000 zlotys. On this basis, a 50 per cent reduction in agricultural employment during the period 1976–90 would require investment outlays of up to 750 billion zlotys.

For the period 1976–90 about the same rate of growth for agricultural production as in the past fifteen years is anticipated, resulting in an absolute rise in output of about 50 per cent. This will require correspondingly higher investment outlays. In the fifteen years 1961–75 gross investment outlays for agriculture exceeded 500 billion zlotys. Consequently, in the coming fifteen-year period investment outlays to increase production potential, together with value of the replacement of used production means, should amount to about 750 billion zlotys. Thus, the gross investment outlays for agricultural production growth and for structural changes in agriculture will total approximately 1500 billion zlotys in the period 1976–90, with the actual magnitude of these outlays lower or higher according to the rate of growth of the whole national economy. This rate will determine, on the one hand, the demand of the non-agricultural sectors for manpower from agriculture and, on the other hand, it will determine the investment possibilities of the national economy.

## DEVELOPMENT PLANNING OF THE COMPLEX OF THE FOOD ECONOMY

So far the planning and economic instruments for directing the development of agriculture as a separate branch of the national economy, covering the production of raw materials and agricultural goods, has been discussed. This approach to the development planning of agriculture is however too narrow to be sufficient. Agriculture does not exist or develop in isolation from other sectors of the national economy. It is a part of the whole economic organism and it has to be linked strongly with it, both by streams of material goods and services flowing from one sector to another (inter-sectoral flows) and by the system of managing and directing production.

Scientific bases are now being formed in Poland for planning the development of the entire food economy, rather than of agriculture only. The term used in this connection is 'the complex of the food economy'. It covers the following elements.

Industries which manufacture the means of production for agriculture and for the food industries, i.e. the machine and equipment industry for agriculture and for the food industry, fertilizer and pesticide industry, building material industry and other sectors and branches of the national economy which provide material and technical conditions for food production. The share of each of these sub-sectors in the complex of the food economy is proportionate to the extent of the flow of its total production to the food-producing branches.

Actual agriculture in the sense of agricultural production excluding, however, those agricultural products that are not connected with food (raw materials for textile or leather industries, etc.).

Agricultural commodity and food processing, the main function of which is processing of agricultural raw materials into final food products.

Services, in a broad sense of the word, concerned with food production and distribution through to the consumer. It also covers the activities required for improvement of qualifications of persons employed in all parts of the food economy.

The complex of the food economy in Poland has no formal organizational structure. It is, however, an important cross-section of the whole national economy. Thus, what integrates these various sectors and subsectors of the national economy is the process of food production in the broad sense. The food economy, referring to the whole range of processes from farm production to final retail purchases, is an extremely important element of Poland's economy. According to the recent studies, it employs 48 per cent of the professionally active persons. It has available 40 per cent of the total value of production assets of the entire national economy and it provides 37 per cent of the domestic total production and 38 per cent of the national income. Thus it is one of the biggest and most important elements of the Polish economy.

The main problem is that all the elements of the complex of the food economy, in which nearly all sectors and branches of the national economy participate in some manner, must be programmed and developed harmoniously. A uniform, internally integrated mechanism for directing this complex is also being formed. These links are very complicated and require a new approach and a new methodology of economic planning. The first complex, long-range programme for the development of the entire food economy during the period 1975–90 was elaborated and approved by the highest State authorities including the Seym (Parliament) in November 1974, and is now being implemented. It provides for a considerable increase of investment outlays for the development of the food economy. It also changes and modernizes its internal structure and eliminates poorly developed links which slow down the development of the whole complex.

Adoption of the concept of the complex of the food economy means thorough changes in the system of economic planning. Agriculture cannot now be planned alone, and consequently the plan for the development of agriculture is prepared and approved as a part of the over-all plan for development of the total food economy. Directing the complex of the food economy involves primarily the following:

shaping the proportions between the extent and rate of development of the food complex and the entire national economy, as well as the needs of foreign trade;
shaping or optimizing the internal structure of the complex, particularly of material proportions between its individual elements (sector-branch structure);
selection and introduction of adequate techniques for food production;
changes of socio-economic structures in the food complex (forms of

ownership of the means of production and the respective social forms of organization of production);

production processes, i.e. shaping the volume and structure of the final product of the complex;

solution of social problems of the employees of the complex (wage and social policy, living and working conditions).

Action in these areas determines simultaneously the content of the food complex plan. It is being implemented by different methods and by using different instruments according to the sector and character of the separate aspects. The complexity of directing the development of the food economy in Poland consists in the fact that there are over 3.5 million economic units, including almost 3.4 million individual producers. An over-all approach to the issues of food production means that, *a priori*, operations promoting a harmonious development of the entire food economy can be foreseen and planned. What is particularly important is the coordination of the programme for growth of agricultural production with the deliveries of the means of production for agriculture and with the production capacities of the food industry, which is the main customer for unprocessed products of agricultural origin.

# 7 The Challenge of the Future

## EMPLOYMENT AND LABOUR PRODUCTIVITY IN AGRICULTURE AND IN NON-AGRICULTURAL SECTORS

In the years 1950–74 the population of Poland increased by 8.8 million persons, i.e. by 35.5 per cent (Table 7.1), while the number of persons employed grew by 6.6 million persons, i.e. by 65 per cent. Employment in relation to the total population, or the labour participation ratio, increased from 41 per cent in 1950 to 49 per cent in 1974. This was primarily the result of women's professional activity in non-agricultural sectors of the national economy.

Employment outside agriculture grew from 4.7 million persons in 1950 to 11.6 million in 1974, i.e. by 147 per cent. The annual employment growth rate averaged 3.8 per cent. Such a high rate of increase of employment was feasible because of the growing industrialization of the country. The scale of the country's industrialization is shown by the fact that between 1950 and 1974 the national income produced outside agriculture increased ninefold while in agriculture it grew by only 50 per cent. In spite of the fast rate of economic development of the country (7.3 per cent yearly in the years 1950–74, the national economy was unable to bring about a substantial reduction of employment in agriculture. This declined by only 13 per cent over the period.

In the period analysed the value of agricultural production (in constant prices) grew on the average by 2.9 per cent yearly, and by 3.4 per cent per employee. In view of the growing capital intensity of agricultural production, labour productivity per employee, measured by the value added,[1] increased more slowly, by only 2.3 per cent annually as compared with 5.6 per cent in the non-agricultural sectors. Consequently the difference in the level of labour productivity continued to widen to the disadvantage of agriculture. Whereas in 1950 labour productivity in agriculture was 62 per cent of that in the non-agricultural sectors, by 1974 it had declined to 28 per cent. The low rate of growth of labour productivity in agriculture reflected the low rate of decline of employment (0.5 per cent yearly) and the difficulties of adjusting individual farming to the requirements of technical advance.

TABLE 7.1 Population, employment, national income and labour productivity in agriculture and in other sectors of the national economy, 1950–74

| | 1950 | 1960 | 1970 | 1974 | Average annual rate of growth 1960/1950 | 1970/1960 | 1974/1970 | 1974/1950 |
|---|---|---|---|---|---|---|---|---|
| | (million persons) | | | | (percentage) | | | |
| Population | 24.8 | 29.6 | 32.5 | 33.6 | 1.8 | 1.0 | 0.8 | 1.1 |
| Employment[a] | | | | | | | | |
| Total | 10.2 | 12.4 | 15.2 | 16.5 | 1.9 | 2.0 | 2.1 | 2.0 |
| In material production | 9.2 | 11.0 | 12.8 | 14.0 | 1.1 | 1.5 | 1.7 | 1.4 |
| in agriculture | 5.5 | 5.4 | 5.2 | 4.9 | 0.2 | 0.4 | 1.5 | 0.5 |
| outside agriculture | 3.7 | 5.6 | 7.6 | 8.8 | 4.2 | 3.1 | 3.8 | 3.7 |
| Total outside agriculture | 4.7 | 7.0 | 10.0 | 11.6 | 4.1 | 3.6 | 3.8 | 3.8 |
| National income[b] | (billion zlotys) | | | | | | | |
| Total | 213 | 441 | 791 | 1157 | 7.6 | 6.0 | 10.0 | 7.3 |
| in agriculture | 102 | 120 | 135 | 154 | 1.6 | 1.2 | 3.3 | 1.7 |
| outside agriculture | 111 | 321 | 656 | 1003 | 11.2 | 7.4 | 11.2 | 9.6 |
| Labour productivity[c] | (thousand zlotys) | | | | | | | |
| Total | 23.2 | 40.1 | 61.8 | 84.5 | 5.6 | 4.5 | 8.1 | 5.6 |
| in agriculture | 18.5 | 22.2 | 26.0 | 31.4 | 2.0 | 1.6 | 4.8 | 2.3 |
| outside agriculture | 30.0 | 57.3 | 86.3 | 114.0 | 6.7 | 4.2 | 7.2 | 3.7 |
| Share of agriculture | (percentage) | | | | | | | |
| In total employment | 54.9 | 43.5 | 34.2 | 29.6 | | | | |
| In employment in material production | 59.8 | 49.1 | 40.6 | 35.8 | | | | |
| In national income | 47.9 | 27.3 | 17.1 | 13.3 | | | | |

[a] In terms of full-time employees.
[b] At 1971 prices.
[c] National income per employee.

The growing disproportion between the growth of labour productivity in agriculture and that in non-agricultural sectors also led to a growing disparity in incomes, to the disadvantage of agriculture. Since the basic principle of state policy is an equal rate of growth for incomes of agricultural and non-agricultural populations (which obviously is not the same as equal levels of income), the equality of income growth rates was brought about by raising real prices for agricultural products. Thus, for example, in the period 1960–74 about 57 per cent of the growth in the real income of individual farms was a result of increased agricultural production while 43 per cent was a result of changes in price levels and relationships in favour of agriculture.

In order to counteract inflation, State policy makes efforts to limit upward trends in living costs, particularly in food expenses.[2] Thus in the five-year period 1971–5 the principle of unchanged prices for basic food articles (cereal products, dairy products, meat and sugar) is obligatory. Under these circumstances, raising the purchasing prices for agricultural products makes it necessary to increase State subsidies for retail food prices.

Further growth of the disproportion between the purchasing prices for agricultural products and retail food prices cannot be maintained in the long run. This dilemma can only be solved by providing such conditions for growth of labour productivity in individual farming as will ensure an adequate growth of incomes for the farm population without necessarily supplementing them by a continual rise of real prices for agricultural products. This can be achieved only by reducing employment in agriculture per unit of land area.

According to projections, up to 1990 the consumption expenditure out of personal incomes per employee in the national economy should grow by about 5 per cent yearly. In order to ensure the same rate of growth for the consumption expenditure in peasant farming without raising real prices for agricultural products, its labour productivity should grow by at least 5.5 per cent annually. This latter rate is required because of the necessary increases in the share of investment in incomes to cover the investment and other means required for increasing agricultural production.

According to estimates of the Institute of Agricultural Economics, in order to ensure that this rate of increase in labour productivity per employee in individual farming (5.5 per cent annually), is attained, employment per unit of land area should decline by 4 per cent per year. The number of full-time employees should decrease from about 28 persons per 100 hectares of agricultural land in 1974 to about 14 to 15 persons in 1990. The area per employee would thereby be increased from 3.6 to about 7 hectares of agricultural land. The average area per family in individual farming should grow from 4.8 to about 12 hectares of agricultural land, and if small part-time farms are excluded, the average area per family in exclusively agricultural farms should expand up to 15 hectares of agricultural land. Such an increase will be achieved both by expanding the size of family farms (individual farms) and by setting up group farms (as will be discussed later). In agricultural cooperatives the area per associated family should grow from 9 hectares in 1973 to about 18 in 1990.

## DEMOGRAPHIC PROJECTIONS AND THE DEMAND FOR MANPOWER

The rate of population growth in Poland has declined steadily from 1.8 per cent yearly in the fifties and 1.0 per cent in the sixties to 0.8 per cent in the seventies. According to demographic projections, in the years 1981–95 the increase of the population will average 0.6 per cent yearly (Table 7.2).

The rate of population growth in the economically productive age group remains fairly high in the present decade; during 1971–5 it was 1.8 per cent and during 1976–80 it will still average 1.4 per cent annually. After 1980 the rate of population increase in the productive age group will decline considerably to average 0.7 per cent annually in the period 1981–5, and 0.4 per cent in 1986–95.

TABLE 7.2 Demographic projections for Poland to 1995

| *Year* | *Total population* (*million*) | *Yearly average rate of growth in 5-year periods* (*percentage*) | *Population at productive ages*[a] | | |
|---|---|---|---|---|---|
| | | | *total total* (*million*) | *increase during 5-year periods* (*million*) | *average yearly rate of growth in 5-year periods* (*percentage*) |
| 1970 | 32.6 | — | 18.3 | — | — |
| 1975 | 33.9 | 0.8 | 20.0 | 1.7 | 1.8 |
| 1980 | 35.3 | 0.8 | 21.4 | 1.4 | 1.4 |
| 1985 | 36.5 | 0.7 | 22.1 | 0.7 | 0.7 |
| 1990 | 37.4 | 0.5 | 22.5 | 0.4 | 0.4 |
| 1995 | 38.5 | 0.6 | 23.0 | 0.5 | 0.4 |

[a] Men aged 18–64, women 18–59.

In the present decade the population increase in the productive age group, in absolute figures, will amount to 3.1 million. The manpower increase will be somewhat smaller in view of the anticipated growing number of students, the possibility of earlier retirement in some professional groups, and the extension of paid maternity leaves and the opportunity to obtain longer unpaid maternity leaves. Besides, in spite of the continuing increase in the professional activity of women, particularly younger ones (at present, outside peasent farming, 63 out of every 100 women of the economically productive age group are professionally active), some women do not take up professional work. This is particularly the case for workers' wives in such occupations as mining, metallurgy and others where, according to tradition, the wife should not be employed outside but should run the household. The power of this tradition is weakening now but is still considerable.

In the past decade employment outside agriculture increased by 3 million persons (from 7 to 10 million), or by 300,000 persons per year. In the first four years of the present decade (1971–4) employment was increasing somewhat more rapidly, by 400,000 persons yearly. This increase in the rate of employment growth results not only from a higher manpower supply (youth of the so-called population explosion period entering the workforce) but also from the acceleration of economic development of the country, which resulted in a more rapid expansion of new working places in industry and in other non-agricultural sectors. Although the growth in employment will be somewhat slower in the coming years, the general employment growth outside agriculture will be much higher in the current decade as a whole than in the previous one.

Since the effective manpower increase in the national economy will be lower than the population increase in the economically productive age group (less than 3 million persons), it will be possible to satisfy the manpower demand of non-agricultural sectors only by a decline of employment in agriculture. During the present decade it is expected that some 6–800,000 persons will leave agriculture, mainly in the five years 1976–80. As compared with 1970, the decrease in employment in agriculture during the decade will however not be very great, only 12 to 15 per cent. Consequently, changes in the agrarian structure in the current decade will have to be rather limited.

In the subsequent decade (1981–90), the population increase in the economically productive age group will amount only to 1.1 million persons for the country as a whole. Under these circumstances, it will be necessary for about 2 million persons to leave agriculture in order to satisfy the manpower demand of non-agricultural sectors. In all, the employment in agriculture would decrease from 5.2 million persons in 1970 to about 4.5 million in 1980 and about 2.5 million in 1990. This would mean an average annual rate of decline of 5.7 per cent during the 1980s. This very high rate of decline may not be attainable, not because of the lack of employment possibilities outside agriculture, but because of the need to maintain a higher employment level in agriculture itself. However, regardless of the actual rate of employment decreases in agriculture, it will be very much higher than in the current decade.

The high rate of decrease in employment must be accompanied by a radical increase in the technical equipment of agriculture and by rapid structural changes. The implementation of these changes will be more difficult in that it must be carried out simultaneously with a further growth

TABLE 7.3 Average yields, livestock numbers and production of major agricultural commodities, 1950–5 to 1971–4

| | *1950–5* | *1956–60* | *1961–5* | *1966–70* | *1971–4* | *1974* | *Average annual rate of growth*[a] *(percentage)* |
|---|---|---|---|---|---|---|---|
| | | | *(100 kg per ha.)* | | | | |
| Average yields | | | | | | | |
| Cereals, total | 12.6 | 15.0 | 17.3 | 20.1 | 25.6 | 28.3 | 3.6 |
| Wheat | 13.1 | 16.1 | 19.7 | 23.2 | 28.2 | 31.7 | 3.9 |
| Potatoes | 117 | 131 | 154 | 176 | 176 | 181 | 2.0 |
| Sugar beet | 187 | 211 | 267 | 324 | 315 | 325 | 2.6 |
| Oilseed crops | 6.5 | 9.5 | 13.9 | 17.5 | 16.8 | 20.1 | 4.9 |
| | | | *(million)* | | | | |
| Livestock numbers | | | | | | | |
| Cattle, total | 7.4 | 8.4 | 9.7 | 10.8 | 12.0 | 13.0 | 2.4 |
| cows | 5.2 | 5.8 | 6.0 | 6.1 | 6.1 | 6.2 | 0.8 |
| Pigs | 9.5 | 11.9 | 13.1 | 14.0 | 18.5 | 21.5 | 3.4 |
| | | | *(million tons)* | | | | |
| Production, total | | | | | | | |
| Cereals | 11.8 | 14.0 | 15.0 | 17.0 | 21.3 | 22.9 | 3.0 |
| Potatoes | 30.8 | 36.3 | 43.6 | 47.9 | 47.2 | 48.6 | 2.1 |
| Sugar beet | 6.5 | 7.7 | 11.4 | 13.6 | 13.4 | 13.0 | 3.7 |
| Milk | 8.9 | 11.8 | 12.5 | 14.2 | 15.5 | 16.3 | 2.8 |
| Meat[b] | 1.2 | 1.6 | 1.9 | 2.2 | 2.6 | 3.1 | 3.9 |
| Production' per caput | | | | | | | |
| Cereals (kg) | 436 | 469 | 471 | 505 | 639 | 680 | 1.9 |
| Potatoes (100 kg) | 11.9 | 12.6 | 14.2 | 14.9 | 14.3 | 14.4 | 0.9 |
| Sugar beet (kg) | 250 | 268 | 372 | 422 | 402 | 385 | 2.4 |
| Milk (litres) | 341 | 391 | 406 | 440 | 466 | 483 | 1.5 |
| Meat (kg)[b] | 45.5 | 55.7 | 57.9 | 62.3 | 79.6 | 91.2 | 2.7 |

[a] 1950–5 to 1971–4.
[b] Post-slaughter weight, including fat and offal.

of agricultural production necessary to satisfy the food demands of the increasing population and to maintain essential exports.

## AGRICULTURAL PRODUCTION AND FOOD CONSUMPTION

The development of agricultural production depends on increases in the yields of cultivated crops, on changes in the area of land used for agriculture and for animal production, and also on increases in fodder imports. During the period 1950–74 the average yield of all crops cultivated in Poland rose by 90 per cent, or by an annual average of 2.7 per cent. The most rapid growth of yields was in oilseeds (4.9 per cent) and cereals (3.6 per cent), while the lowest growth was in potatoes (only 2 per cent yearly). Since the agricultural area decreased in those years by about 5.5 per cent (from 20.4 to 19.3 million hectares), the total output of all cultivated crops rose by 2.5 per cent annually. It was possible to maintain the average yearly rate of growth in yields at 2.7 per cent by increased fertilization (from 17.7 kg NPK in pure elements per hectare of agricultural land for 1950 crops to 173.6 kg NPK for 1974 crops) and by the introduction of new and more intensive varieties, as well as by the use of pesticides to fight pests and plant diseases (the demand for the latter in Poland is, in general, fully satisfied). (See Table 7.3.)

During this period, animal production grew faster than crop production, by 3.1 per cent yearly. Higher rates of growth of animal production were possible owing to an increased share of the crop production being used for feeding of productive livestock (a decreased share was used for food, seeds and fodder for horses) and owing to increased imports of cereals and high-protein feeding stuffs. These imports increased from about 0.6 million tons annually in the period 1950–5 to about 4 million tons annually in the period 1973–4.

The increase of the gross value of production of agriculture[3] in the period under study at constant 1971 prices is as follows:

| *Year* | *Total (billion zlotys)* | *Per caput (thousand zlotys)* |
|---|---|---|
| 1950 | 125.0 | 5.0 |
| 1960 | 155.8 | 5.3 |
| 1970 | 198.3 | 6.1½ |
| 1974 | 246.6 | 7.3 |

The average annual rate of growth of agricultural production was as follows:

| *Period* | *Total* | *Per caput* |
|---|---|---|
| | *(percentage)* | |
| 1951–60 | 2.3 | 0.6 |
| 1961–70 | 2.4 | 1.4 |
| 1971–4 | 5.6 | 4.6 |
| 1951–74 | 2.9 | 1.6 |

Polish agriculture has been and still is under strong pressure from the demand for food. The strong demand for food in the domestic market is the result of socio-economic changes that have brought rising general living standards for the population. Large numbers of low-income people with increases in their income desire first to raise their food consumption. Moreover, it is the result of an absolute growth of demand for food following the faster migration from agriculture to non-agricultural sectors. A basic factor is that the pressure of demand on the food market is further strengthened by a still relatively inadequate supply of industrial products, particularly consumer durables, which is typical for periods in which the capital goods industry develops first.

The role of food in the development of the national economy and the consequent tasks for agriculture are determined by the continuing high share of expenditure for food in total consumer expenditure (50 per cent in 1955, 49 per cent in 1965, 43.7 per cent in 1970 and 40.8 per cent in 1973). This is the result of the rapid rate of growth of incomes and the high income and price elasticity of the demand for food. All these factors, together with the role of food exports in the balance of payments, make the balance in the domestic market and even the general economic equilibrium depend essentially on the situation in agriculture. Any significant drop in the level or slowing down of the rate of growth of agricultural production causes economic difficulties.

Generally speaking, Polish agriculture, in spite of some fluctuations in the production level, successfully implements its tasks in this field, as is shown primarily by the evolution of food consumption. From the viewpoint of an international comparison the present nutrition level in Poland should be considered as sufficient. The level of food consumption, especially of animal protein, has grown significantly, particularly in the period 1971–4 (Table 7.4). A comparison of food consumption in Poland with that in developed European countries indicates that the difference between Poland and those countries is fairly small and much smaller than the difference in the levels of national income per caput.

However, in spite of the higher level and improved structure of consumption, improvements are still needed. The changes that have taken place have not sufficiently influenced the nutritional value of the daily diet.

Acceleration of these changes is one of the major tasks of agricultural policy in the near future. In the projections of the country's economic development up to 1990, an increase of meat consumption to about 99 kg yearly

TABLE 7.4 Per caput consumption of selected agricultural products, 1950–74 and projections for 1990

| *Product* | *1950* | *1960* | *1965* | *1970* (*kg per year*) | *1973* | *1974* | *1990*[a] |
|---|---|---|---|---|---|---|---|
| Cereals[b] | 166 | 145 | 141 | 131 | 125 | 123 | 90 |
| Potatoes | 270 | 223 | 215 | 190 | 183 | 177 | 110 |
| Meat | 36.5 | 42.5 | 49.2 | 53.0 | 62.1 | 65.5 | 85–90 |
| Fish and fish products[c] | 1.7 | 4.5 | 5.0 | 6.3 | 7.2 | 7.3 | 11–12 |
| Edible fats, total[d] | 9.7 | 13.6 | 15.1 | 18.0 | 19.3 | 19.6 | 26 |
| animal fats | 5.5 | 6.6 | 6.1 | 7.3 | 7.6 | 7.5 | 7.5 |
| vegetable fats | 1.5 | 3.1 | 4.9 | 5.7 | 6.2 | 6.5 | 8.5 |
| butter | 2.7 | 3.9 | 4.1 | 5.0 | 5.5 | 5.6 | 10 |
| Milk and milk products (litres)[e] | 206 | 227 | 233 | 262 | 263 | 263 | 340 |
| Hen's eggs (units) | 116 | 143 | 162 | 186 | 200 | 205 | 270 |
| Sugar | 21.0 | 27.0 | 32.6 | 39.2 | 41.6 | 44.0 | 49 |

[a] *XV Plenum of the Central Committee of the Polish United Workers Party*, Warsaw, 1974, p. 20.
[b] In terms of processed articles.
[c] In trade weight.
[d] In terms of pure fat.
[e] Without milk processed into butter.

per caput, or from about 2.2 million tons in 1974 to about 3.4 million tons in 1990, is envisaged. Thus, meat consumption should keep growing by about 2.9 per cent per year. Total production of cattle for slaughter should grow from 4.2 million tons live weight in 1974 to over 7 million tons in 1990, or by about 3.3 per cent annually. The assumed growth of cattle production covers not only the satisfaction of the needs of the domestic market but also exports of meat and meat products.

The rate of growth of cattle production will continue to be faster than the rate of growth of total agricultural production, which over the period 1975–90 should amount to about 3.1 per cent a year on average. The value of final production in constant 1971 prices should grow from about 247 billion zlotys in 1974 to over 400 billion in 1990.

Thus, the projected rate of growth of agricultural production till 1990 is somewhat higher than that achieved in the period 1950–74 (2.9 per cent yearly). Achieving this rate with a 50 per cent decline of employment in agriculture and with rapid structural changes will be a difficult task not only for agriculture but for the entire national economy, which has to increase very greatly the investment outlays for agriculture and to supply agriculture with current production means such as fertilizers, industrial fodder, fuel, etc.

## MAIN PREMISES OF THE AGRICULTURAL RECONSTRUCTION POLICY

The main trend and strategic target of transformations of Polish agriculture is determined by the socialist character of the country. This target is to change small-scale family farming into large-scale socialist farming. The experience of other socialist countries, which have already accomplished this transformation, indicates that socialization of the means of production in agriculture provides the starting point for further social, technical and organizational changes. The aim of the change is to adjust agriculture to new levels of technology and thus to create conditions for increased labour productivity and for improved effectiveness of farming. At the same time the aim is to ensure working and living conditions for the agricultural population similar to those in other sectors. These aims are implemented by gradual introduction of industrial production methods and organization, brought about by both permanent quantitative and qualitative increases of the technical equipment for agriculture, and by further concentration of production in larger-scale organizational units.

Present trends and directions of the development of agriculture in socialist countries indicate the shape of agriculture at the turn of the current century. The implementation of the future model of agriculture is not a simple issue. This is not only because of the magnitude of the future inputs and outputs involved but also because of the necessary experimentation and searching for optimum solutions, particularly with respect to the socially, economically, technically and ecologically rational concentration of production.

The experiences of socialist countries at the stage of post-collectivization in agriculture, especially in setting up new organizational forms, have by now been drawn upon in Poland, mainly in the State sector of agriculture, in the organization of multi-unit agricultural enterprises and agricultural – industrial combines, and in the introduction of industrial methods in agricultural production. Though the State sector in Polish agriculture presently covers only one-sixth of the agricultural land, new organizational forms are being introduced on a considerable part of this area, which is concentrated in compact complexes, in order to work out models for the future organization of agricultural production. Examples from the socialist countries indicate that the organizational forms in agriculture do not depend so much on the character of ownership (State or cooperative) as on the advance of technique and production technology. Therefore, the organizational forms of agricultural production being worked out now by State farms will be applied in the future in the agricultural sector as a whole. Thus the significance of the State sector in Polish agriculture goes far beyond its present quantitative share in the country's agricultural production.

Although the share of the State sector continues to grow year by year, owing to the take-over of land transferred to the State by farmers in exchange for pensions, this form of socialized agriculture will have a limited scope. The fundamental process of reconstruction of agriculture will take place within family farming, with the voluntary active participation of farmers themselves. Thus, in Polish agriculture there will occur two parallel processes: on one hand, the latest organizational production forms in the socialized sector will be developed and improved, while on the other hand small-scale farming will be gradually transformed into larger group or cooperative enterprises.

The level of economic development that has already been achieved in Poland provides incomparably greater opportunities for social and technical transformation of agriculture than those available in the early fifties. In the past twenty-five years, the situation of Polish agriculture has changed fundamentally. Agricultural production has almost doubled and the equipment of agriculture with production means, particularly modern technical means, has increased greatly. This is confirmed for example by the number of tractors, which has risen from about 20,000 in 1950 to over 360,000 in 1974. Also the ability of the national economy to provide investments has increased many times. The investment outlays for agriculture in the five-year period from 1976 to 1980 will be ten times higher, and in the subsequent five-year periods fifteen and twenty times higher, than in the period 1951–5.

The need for constant growth of agricultural production requires a maximum use of the production potential in all sectors of agriculture, including the peasant sector. The latter will still determine the general level of agricultural production in the country for a long time to come. Regardless of the advancing socialization of agriculture, simultaneous structural changes within individual farming are necessary not only for purely economic reasons (that is, for ensuring the increase of labour productivity) but also for creating conditions for further changes in agriculture. These depend, among other things, on reducing the number of families employed in farming. In the future large-scale farming, employment will be found by one-third, at the most, of the present number of individual farmers. The setting up of socialized farms at the present employment level would not be economically rational, both because labour resources would not be used adequately and because of the need to ensure an appropriate income level for the agricultural population.

The agricultural policy of gradual structural changes spread over a longer period of time aims, on one hand, at securing conditions for constant production growth within individual farming and, on the other hand, at enabling the operating and newly established socialized farms to achieve a production level per unit of land area higher than in individual farming. The extent to which these aims are attained will determine the rate of growth in Poland's total agricultural production with the increasing share of socialized farming.

Efforts to ensure the growth of agricultural production at a time of socio-economic changes are realistic, provided that these changes are preceded by an adequate increase in the technical equipment of agriculture. The technical reconstruction of agriculture, and the outflow of manpower connected with it, must be a condition for, rather than a result of, transformations of the system. The extent to which advances can be made simultaneously in the socialization of agriculture and in maintaining the growth in agricultural production is conditioned by the possibilities of the national economy for supplying agriculture with production means and by its capacity to absorb manpower released from agriculture.

## LAND CONCENTRATION

The area of agricultural land decreased in Poland from 20.4m hectares in 1950 to 19.3m hectares in 1974. It is expected that this area will decline to about 18.6m hectares by 1990. The decrease of the area of agricultural land results both from some of it being taken over for construction of industry, towns, roads and water reservoirs, and for afforestation. About 12 per cent of the agriculturally used soils in Poland are poor sandy types. Under the pre-war conditions of agrarian overpopulation and 'land hunger', all land, including even the poorest, was utilized. Today, farming on this marginal land becomes ever less profitable and, therefore, a part of the poorest land is designated for afforestation.

The area of agricultural land per inhabitant amounted to 0.82 hectares in 1950, 0.57 in 1974, and it will drop to about 0.50 hectares by 1990. The decreasing area of agricultural land per caput and the necessary growth of agricultural production by 1990 of about 65 per cent in comparison with 1970 calls for effective use of all agriculturally productive land. A prerequisite for decreasing employment in agriculture and for the inflow of manpower to non-agricultural sectors is land concentration in larger production units. The forms of land concentration preferred by agricultural policy are those which ensure its best intensive use.

Land concentration in agriculture is implemented along two parallel lines:

- the voluntary merging of land by individual farmers for common farming;
- the winding up of a part of the individual farms and the taking over of this land by other users.

At present, over 3.4 million families in all sectors are linked with agriculture, including 2.4 million families engaged full time and one million part time. It is assumed that by 1990 there should remain in agriculture no more than 1.5 million families, of which about one million would be families employed in agricultural production only and about 500,000 would be dual-occupation families. Nearly two million families should therefore leave agriculture. Of these about one million will still have their residence in the village but will give up operation of a farm and keep for their own use only a small plot. The other million will mostly migrate to urban areas. One of the tasks of economic policy is to bring about a polarization of dual-occupation farms, in the sense of either reducing the farm with retention of only a small plot or of expanding the area of the farm with the setting up of a typically agricultural farm.

At present, over one-third of the individual farm owners are aged 60 and over. Most of these have no successors to take over their farms. These farmers when reaching the age of 65 (for women, 60) may transfer land to the state against a pension. This form of transferring land has existed since 1968. It was not, however, too attractive for most of the farmers. In order to accelerate land turnover and to encourage farmers to make earlier transfer of their land to the State (farmers often continue to operate their farms at the age of 70 or over), a law was issued in 1974, 'On transferring farms for State ownership against pension or cash payments'. According to previous regulations, a farmer was entitled to a pension after he transferred to the State at least 5 hectares of agricultural land. The new law lowered this minimum to 2 hectares. The amount of the pension depends on the number of transferred hectares of agricultural land; for the first 2 hectares, 300 zlotys monthly for each hectare; for the next 8, 100 zlotys monthly for each; and for every hectare over 10, 50 zlotys monthly. If the farm includes forest, the pension is increased by 100 to 300 zlotys monthly according to the value of the stand of trees. If the farmer transfers buildings together with the land, the pension is also increased from 100 to 300 zlotys monthly according to the value of the buildings. If the farmer renounces his right to living quarters free of charge, the pension is also raised by 150 zlotys monthly. In all, a monthly pension for the farm transferred to the State cannot, however, exceed the amount of 2500 zlotys, which was about 80 per cent of the average worker's wage in the national economy in 1974.

Farmers who do not meet the conditions for obtaining a pension (for example, if they have less than 2 hectares of land) or who already are receiving pensions (for example peasant-workers), can transfer land to the State for cash at prices determined in land transfers between the State and farmers. These prices also apply to the sale of land to farmers by the State. The farmer receives for the land sold to the State 50,000 zlotys in cash and the remaining part in yearly instalments, bearing 3 per cent interest.

Recently, the State has been taking over from individual farmers over 100,000 hectares yearly. It is anticipated that by the end of the 1970s this figure will be doubled. About 70,000 hectares are transferred to State farms and the remaining land to other users, including individual farms. In all, individual farms are using over 600,000 hectares of land under lease from the State Land Fund.

In addition to the land turnover through the State Land Fund, there are also land transfers among individual farmers. Group and cooperative farms, and recently State farms, can also buy land directly from private farmers. The State supports private land transfers by granting easy credits for this purpose, particularly to young farmers who are buying whole farms.

One of the conditions to ensure adequate growth of labour productivity in small-scale farming is the necessary increase of land area per family, either by an individual augmenting the size of his farm or by the taking over of land for common farming by a group of individual farmers. Therefore, in the regions of dispersed agrarian structure, it is preferred to distribute land from the State Land Fund, to improve agrarian structures within small-scale farming. In these regions, land from the State Land Fund is transferred, to the extent feasible, to groups of farmers for use on long-term leases, and otherwise to individual farmers in the form of sale or long-term lease.

In some regions of the country with a relatively more favourable agrarian structure and consequently with smaller demand for land on the part of individual farmers (mainly in western and northern voivodships), one of the main forms of bringing vacated land under cultivation is for State farms to take it over. Where the vacated land is not adjacent to the State farm area or where the complexes of vacated land do not allow a rational State farm to be organized, other forms of bringing land under cultivation are used, including agricultural cooperatives or farmers' groups and individual farms.

## CHANGES IN INDIVIDUAL FARMING

A most essential problem of Polish agriculture for the near future is the adjustment of family farming to the demands of the fast-growing national economy. Family farming has so far performed successfully not only in terms of production but also in providing employment and the maintenance of a considerable part of the country's population at fairly low social costs. High employment per unit of land area and low technical equipment for these farms played a favourable part in getting the country out of economic backwardness. In the conditions of a developed economy, these advantages of family farming are becoming increasingly its weak point. Family farming in Poland is now at a turning point in its historic development. There appear ever more sharply technical and economic barriers in its development in the present shape. Because of the small size of these farms and their small scale of production, effective application of mechanization and of technical installations are not possible. The economic barriers of family farming are reflected in the inadequate family income and in the difficulties in ensuring socially adequate growth of this income from the unaltered farm area.

Both a considerable decline of family farming and the insufficient supply of tractors and agricultural machines provided the basis for setting up and developing group service forms of mechanization in Polish agriculture (concentration of agricultural equipment in agricultural circles). The aim was to ensure at least a minimum of mechanized services to all farms that needed them. The service form is still being developed, but it becomes ever more insufficient. It is unable to solve all the mechanization problems, particularly in bigger farms which aspire to have available their own tractors and machines. In addition to service mechanization, there also develops individual mechanization, as shown by the following data:

*Number of tractors*

| | *Agricultural circles* | *Individual farms* |
|---|---|---|
| | *(thousands)* | |
| 1960 | 6.9 | 13.3 |
| 1965 | 45.6 | 23.0 |
| 1970 | 91.9 | 54.6 |
| 1974 | 123.2 | 153.7 |

At present, there are available 1.8 tractors for each 100 hectares of agricultural land in family farming. It is estimated that, to cover the demand fully, at least 5 tractors per 100 hectares are needed. It is assumed that, as in the years 1971–4, individual mechanization will supplement the service activity of agricultural circles. These latter will render mechanization services especially to smallholdings, for which the purchase of their own tractor is not economically warranted.

The increase in the number of tractors in agriculture also contributes to a decrease in the number of horses. On State farms and agricultural cooperatives there are only 1.7 horses per 100 hectares. It appears in practice that if farmers have their own tractors they often give up keeping horses altogether. At present, 14 per cent of the agricultural land of Poland is used to provide feedstuffs for horses on family farms; this amounts to about 2.2 million hectares for the country as a whole. Reduction of the number of horses will be one more source of further growth in animal production to fulfil demand for food.

In addition to tractor mechanization there is an ever-growing demand by farmers for higher-degree mechanization, chiefly for combines and machinery for harvesting grains, sugar beet, potatoes and green forage. It is anticipated that this form of mechanization on family farms will become exclusively a service provided by agricultural circles. In 1974 agricultural circles owned 6300 grain combines. About 8 per cent of the grain area in individual farms is harvested by means of combines.[4] According to forecasts, combine grain-harvesting in individual farms should amount to about 70 per cent in 1985, and to about 90 per cent in 1990.

Whereas mechanization in crop production makes fairly rapid progress, mechanization in animal production is a much more difficult problem to solve. Technical equipment is used for giving fodder, milking or removal of manure on only a small number of the larger farms with large-scale animal production. In an overwhelming number of farms animal production is based almost exclusively on manual work. In this respect the outlook is less optimistic than in the case of crop production. On average, there are about 8 pigs and 4.1 cattle, including 2.3 cows available per family farm (excluding smallholdings below 2 hectares). In such small-scale production the application of mechanical equipment is, naturally, uneconomic. This small production scale also means that the amount of labour per unit of animal production, and consequently the production costs, are high and exceed the price level, particularly on smaller farms. The only way to enable application of mechanization in animal production and the lowering of labour requirements and costs is to increase the scale of production; but this requires a larger farm area.

Expansion of the area of individual farms brings, however, an unfavourable effect, namely a decline in the intensity of farming. This is expressed in lower financial outlays and lower labour input per hectare and, as a result, in lower production and smaller income per hectare but in higher labour productivity per employee. The decrease in intensity of farming with the increase in area of individual farms results particularly from capital shortages. This unfavourable phenomenon is eliminated by increased technical equipment for these farms and by the specialization of production. Specialization of production, animal production in particular, requires intensive farm operation, high capital equipment and considerable cash outlays.

Family farms in Poland are traditionally multi-purpose, except for suburban vegetable farms. The process of farm specialization, particularly in animal production, was initiated on a broad scale only a few years ago. The productive and economic effects of farms specializing in cattle and pig production are, according to studies of the Institute of Agricultural Economics, most encouraging.

The level of agricultural production, particularly net production per hectare of agricultural land, for specialized farms under study does not show the downward trend, with the growth of farm area, that is noted for multi-purpose individual farms. On the other hand, the higher product on smaller farms specializing in pig production is simply a result of the fact that production does not depend on farm size but is based mainly on purchased fodder, as shown by high deficit grain balances for these farms.

The specialized farms under study already in 1972 exceeded the index of production per hectare projected for 1990 for Polish agriculture as a whole. Labour productivity, measured by the value of net output, is some 50 per cent higher in farms of about 11 to 12 hectares and 70 to 80 per cent higher in farms of over 15 hectares as compared with similarly sized groups of multi-purpose farms. These results are achieved despite a much higher employment per unit of area.[5]

Farms of specialized production are primarily larger farms which can ensure an adequate scale of production on the basis of their own fodder resources. These requirements are met, in principle, by farms of about 15 or more hectares of agricultural land. On such farms, both the production level per unit of area and the scale of animal production ensure, on one hand, high labour productivity (about 70,000 zlotys per employee) and low production costs (about 65 zlotys per 100 zlotys of production value) and, on the other hand, available balances of cereals and concentrated feedstuffs.

Smaller farms with an area of up to 10 hectares which specialize in cattle production cannot achieve such scale of production as would guarantee income per employee as high as on the larger farms, in spite of their adverse grain balance which ensures higher animal production per unit of land area. Farms of the same size specializing in pig production achieve, indeed, a scale of production guaranteeing a higher income per employee than those specializing in cattle production, but only owing to large purchases of fodder from outside.

Stimulation by economic policy of the process of increasing farm acreage and of production intensification through specialization will provide a most effective way to maintain an adequate income level in agriculture. Apart from productive and economic advantages, the development of specialization in individual farms also accelerates the creation of conditions favourable for gradual socio-economic changes in agriculture. Intensive and highly profitable farms reach sooner the barrier of social working conditions which can be improved only by more or less close forms of cooperation or integration with other farms. Studies of the Institute of Agricultural Economics show that members of group farms come mostly from among farmers who formerly operated larger, more intensive and specialized farms as compared with average farms in the region concerned.

The rise of incomes and improvement of living and working conditions in non-agricultural sectors arouse similar aspirations among the agricultural population. These aspirations are realized, in the first place, by intensification of agricultural production, which is possible owing to an increased stream of outlays from outside. However, rising incomes require not only constantly higher production levels but also an expanded scale of production. Thus, the aspirations of individual farms to keep pace with the non-agricultural population in raising income and improving living standards provide a strong incentive to intensify and expand production scale.

However, with growing incomes, aspirations arise not only to improve the living standards but also the working conditions, which means regular working hours, days off, holidays, weekends (particularly in view of gradual introduction of a 5-day working week in non-agricultural sectors). Since these working conditions cannot in practice be reconciled with individual operation of a farm, the wish to attain them provides or will provide an essential precondition for changing the forms of farming.

Consequently, the sooner the technical equipment of farms increases, the sooner the agrarian structure within individual farming improves; the sooner the production and incomes of farms grow, the faster will conditions be created for developing new forms of farming leading to gradual socialist changes in agriculture. Such is the logic of structural agricultural adjustment in an economically advanced socialist country. These adjustments should accompany development of small-scale farms and should result from overcoming the technical, economic and social growth barriers by these farms. Therefore, agricultural policy in Poland, by stimulating technical reconstruction and development of agricultural production in small-scale farming, as well as improvement of agrarian structure, is a policy favourable to the creation of conditions for gradual, socialist changes in agriculture.

## FUTURE-ORIENTED AGRICULTURAL ORGANIZATION

Regardless of the length of period of socio-economic transformations in agriculture and of the forms and stages of these transformations, determination of the final agricultural model in the predictable future is of essential significance for current policy as well. The existence of such a model already allows certain processes to be influenced and their development in the desirable direction stimulated. This refers mainly to the spatial management, localization of certain investments projects, the settlement network and the infrastructure.

As noted above, the observation of development trends in socialist countries with fully socialized agriculture gives an idea of the future shape of agriculture in Poland. The various organizational forms being set up at present, though experimental and searching for optimum solutions, indicate that the future model of agriculture will be characterized by agricultural enterprises of large areas and a high degree of production concentration, based on industrial production methods and the organization of production processes.

In all socialist countries, Poland included, there are being established large economic organisms in the form of agrarian – industrial complexes which include agricultural and service enterprises as well as agricultural processing plants. The basic organizational units of these organisms, which naturally have to be limited in area, will not under Polish conditions exceed the area of an existing community, i.e. an average of about 8000 hectares of agriculturally productive land.

The setting up of large economic organisms associating a considerable number of basic organizational production units is, or will be in the future, justified not only by the integration of agriculture and agricultural industry but also by the application of industrial methods in animal production, which allows an unusually high concentration of animals in one production centre. This refers, in particular, to animals which do not require a large storage area for bulky feed, i.e. pigs and poultry.

Assuming that in the year 2000 the number of pigs per 100 hectares of agriculturally productive land in Poland will be 180 to 200, there would be on average 14–16,000 pigs per community. If industrial fattening-houses concentrating some ten or more thousand heads, such as are being set up now mainly in countries with a fully socialized agriculture, prove to be suitable for general use, then under Polish conditions one such fattening-house would serve several communities.[6]

In the case of breeding cattle, which depend to a high degree on fodder production on the farm itself, the animal concentration in comparison with pigs and poultry will be of a much more limited scope. In the future there will be over 3000 head of milch cows and over 4000 head of other cattle per community. In this case, concentrations covering several community production units is out of the question. On the contrary, within a community there will have to be several cattle-breeding farms, taking into account the cost of transporting bulky feed.

If it is assumed that a future, basic agricultural enterprise will cover the area of a present-day community, then present and future changes in agriculture should take into account the organization of such enterprises. At the present stage of agricultural development, with the predominance of individual farming, the problem consists of the gradual inclusion of this farming into the framework of socialized farming. This can take place either directly or indirectly through an ever-closer cooperation and then integration with socialized farming. The integrator of individual farms may, in some regions, be the State farms, but also may sometimes be agricultural cooperatives. In regions where the share of the socialized sector is small, the essential role in the integration of individual farming might be played by an organization which would introduce the future production unit that covered in its scope the one community. The existing community cooperatives of agricultural circles could be such on initiating organization, as in the future these cooperatives will gradually change into agricultural enterprises.

Such an enterprise will include the following production units: specialized farms for animal production, industrial feedstuff plants, fodder drying plants, repair bases, building and improvement plants, and processing plants of a local character for some agricultural products (e.g. vegetables, fruit, milk, breweries, starchworks). It is evident that the functions of such an enterprise can, while becoming consolidated, only gradually cover an ever-wider range. The present initial function of community cooperatives of agricultural circles is mainly the service function of rendering technical services to agriculture. The production function of organizing and operating agricultural enterprises on land vacated by individual farms will however become more significant.

With the consolidation of the cooperatives of agricultural circles as enterprises, the latter will be able gradually to take over new functions integrating the entire range of operations connected with farming in the community area. All types of farms, i.e. cooperative, circle, group and individual, will have an opportunity to develop within such an enterprise. A close production cooperation will then be feasible between the different forms of farming on the basis of specialization in single production phases, particularly in animal production. The cooperatives of agricultural circles will gradually be able to operate not only in the capacity of investors and integrators but also that of shareholders of individual or group farms by direct participation in the financing of their investments.

Individual farms meeting the necessary conditions will have an opportunity to develop and to increase their scale of production without losing future prospects and without uncertainty as to the future, which hampers the natural trend to advance and to expand production. An individual farmer will be able to produce in the framework of a socialist enterprise as long as he can and wants to do so. Inclusion in this way of individual production in the organizational system of a socialist enterprise and of a gradual change from the individual to the group form has the advantage of allowing the old production base and the marginal family labour to continue to be used. This provides not only a theoretical but also a practical opportunity of an evolutionary transformation of individual farming into socialized farming.

## NOTES

1 Value of agricultural production minus value of material inputs.

2 In the years 1960–70 retail prices for commodities and consumption services purchased by the population grew on the average by 1.3 per cent yearly, and in the years 1970–3 by 0.1 per cent yearly (*1974 Concise Statistical Yearbook*, Warsaw, 1974. Main Office of Statistics, p. 641).

3 Including the value of marketed production, self-supply of the agricultural population and increases of the value of livestock and stocks; excluding fodder and seeds grown and used on farms.

4 Agricultural circles harvest, moreover, about 33 per cent of the grain area of individual farms with tractor sheaf-binders.

5 The reduction of employment in these farms with their growing technical

equipment will, of course, bring about an increase of material outlays and a decline of net production per hectare of agricultural land. These changes should however encourage a further growth of labour productivity.

6 In addition to the economic effectiveness which theoretically should be higher for larger concentrations of production, one of the evaluation criteria will be the possibility of checking the unfavourable impact of these farms on the natural environment. The main point is cleaning and utilization of waste matter, and this problem has not so far been solved properly. Therefore, the optimum or maximum size of the pig production concentration in one production centre remains an open question.

# 8 Lessons from Experience

Experience gained in one country can be transferred to other countries only with difficulty and to a limited degree, particularly if the countries differ in their socio-economic and political systems. Polish experiences have been gained under the circumstances of a socialist system, although one in which individual farming continues to predominate. Thus, they result directly from assumptions of a socialist system, particularly the social ownership of basic means of production, the distribution of income according to criteria of social interest and social planning in the broadest sense. There are, however, in these experiences certain elements which may be found relevant for other countries, especially for developing countries which link their development with socialism and set up bases for a planned economy. The following are the major lessons drawn by Polish authorities regarding the experiences of their country.

1 The first and essential condition for accelerating the development of agriculture and integrating it into the national economy is a radical *land reform* – elimination of the remnants of the feudal system, distribution of the land among those working on it, provision of adequate conditions for these people and incentives for them to increase production. (In Poland the land reform was carried out without compensation for estate owners.) The land reform provides initial economic conditions for the development of agriculture. At the same time, it has an essential political significance, as the new authority gains the strong support of the mass of farmers and of the landless rural population.

2 While implementing the land reform the aim should be to set up, according to the conditions, a smaller or larger *State sector in agriculture*. This is not only the beginning of the future socialized farming but, for quite some time, can provide a major support for the modernization of agriculture – for example by promoting production of high-quality seeds and pedigree animals, by widespread dissemination of modern agro-techniques and machinery, by the training of staff for agriculture, and by gaining experience in managing large agriculture enterprises.

3 *The solution of the problem of the economic surplus and of the accumulation of production means* is of decisive significance for the development strategy of agriculture and of the national economy. The success of the development concept adopted in Poland is a most convincing proof that countries

lagging behind economically can, largely on their own, overcome this retardation and break the 'vicious circle of backwardness', provided they solve the problem of accumulating and using the economic surplus in an appropriate and fruitful way. The problems include mobilizing the surplus, accumulating it in the case of a socialist economy in the hands of the State, and designating it for strategic targets. For implementing this course of action, certain preliminary conditions must be fulfilled, such as establishing progressive socio-political relations, ensuring the leading role of the State in the economy, particularly in accumulating the economic surplus and disposing of it, having available central economic planning competence, and setting up social mechanisms which allow the production capacities of the nation to be used.

4 According to the experiences of Poland and other socialist countries, the main *forces and factors influencing agriculture are outside agriculture, particularly in the development of industry*. The country's industrialization finally determines the situation and development trends of agriculture and, in particular, the possibilities and the rate of its technical modernization. Thus at the first stage the strategic target is the development of industry, to which agriculture must contribute by accumulation of capital means, manpower, foreign currency coming from agricultural and food exports, etc. Though the accumulation abilities of a poorly developed and technically primitive agriculture are at first very small, and though they are not sufficient to allow agriculture to finance its own advance, the interests of the total economy including agriculture require acceptance, at the first stage, of the development of industry as the major strategic trend.

5 *The forms of the participation of agriculture in the industrialization of the country and in its total economic development may be various*. In Poland there developed the system of compulsory deliveries of agricultural products, for which the State paid farmers a price 30 to 40 per cent lower than the market price. These deliveries constituted a form of taxation on agriculture for the sake of the national economy. Compulsory deliveries ensured a defined accumulation contribution and, at the same time, they allowed a concentration in the hands of the State of the basic supplies of food items, which are of decisive importance for maintaining balance on the market. Polish experiences show, however, that the extent to which agriculture contributes to the costs of industrialization cannot be made too great because then both the output potential of agriculture and its accumulation capacities decline. Besides the cumulative contribution, agriculture makes its contribution to the over-all economic development by providing manpower for the growing industry, transport, trade, etc. The migration of people from villages to towns and from agriculture to industry creates many complicated economic and social problems, which must be solved in the light of local conditions and needs. At the initial stage of accelerated development there should be allowed a large-scale 'shuttle migration', as a

result of which a considerable group of dual-occupation holders comes into being.

6 As long as there exists, as in Poland, a dominating individual sector in farming, *the linking of individual farmers' means (individual accumulations) with the State means* designated for promotion of agriculture becomes extremely important. This refers most of all to mechanization. Individual mechanization becomes at a certain stage of development unprofitable both from individual and social viewpoints because of its small scale. The State should therefore promote forms of group mechanization. In Poland, the agricultural circles operating on the basis of the Fund of Agricultural Development became the basic form of technical servicing of individual farming and of setting up of group mechanization. It should be emphasized that this form allows the means of individual farmers to be mobilized and linked with the State means. Besides their production purposes, the agricultural circles with their social accumulation also perform definite structural functions by facilitating the future reconstruction of individual farming.

7 The Polish experience indicates that various *forms of production cooperation in agriculture* for overcoming the barrier of small-scale production in individual farming and for raising the effectiveness of outlays in this farming system are of great significance. These forms include both the horizontal cooperation between farmers (groups of individual farmers) and the vertical integration of single farmers and their groups with units of socialized farming, particularly with the agricultural and food industry, with purchase organizations for agricultural products, State farms, etc. Development of these forms also has an essential structural significance because it provides a basis for setting up a cooperative agriculture, in which, although land remains individual property, other production means which play an ever-growing role in intensifying agricultural production become socialized. According to Polish experiences, the development of horizontal and vertical cooperation is a major stage in the transition from traditional individual farming to modern socialist farming.

8 At the present stage of Polish agricultural development the most effective way of improving the agrarian structure is *the setting up of agricultural cooperatives*, which associate the farmers (in terms of voluntary membership) in production, economic and social respects. A cooperative based on group ownership of production means is the best way of linking the interests of the individual farmer with social purposes, in the broad sense. It increases the scale of production, raises its effectiveness, and also helps to solve social and economic problems of the cooperative members. According to Polish experience, introducing the cooperative system in agriculture is a long historical process. It is justifiable to initiate this process by establishing simple forms of cooperation and then to move gradually

from lower to higher cooperative forms. It is extremely important that the organizational forms adopted for the cooperative and the degree of socialization of production means are approved by the farmers.

9 In the Polish conditions a significant role in the development of agriculture was played by *common rural cooperatives of supply and sale*, by credit cooperatives, as well as by some *specialist cooperatives* associating producers of milk, fruit and vegetables, root crops, etc. These form an organizational framework for the development of a wide self-government movement in rural districts. They activate rural environments and stimulate their social and economic activity and facilitate the inclusion of individual farming into the system of social planning. Moreover, they are a major form of spreading widely technical advances in individual farming.

10 During the past thirty years Poland has gained considerable practical experience in *planning the development of agriculture*, particularly in including individual agriculture in the central plan. It is an original achievement, as Poland is one of the few socialist countries in which the sector of individual farming is maintained on a large scale. There have been developed in Poland forms of indirect state influence on individual farming such as, primarily: prices, credit system, contract-buying, and many-sided integration links of individual farming with the cooperatives and the State sector. The effectiveness of the specific forms of planned influence on agriculture is high in Poland, both in short- and in long-term applications.

# III Case Study of Sweden

## CONTENTS

# 9 The Swedish Setting for Agricultural Adjustment

Sweden, with a total area of 45 million hectares, is the third largest country in Europe. Forests cover more than half of its area, however, and an additional 40 per cent consists of mountains, moorlands and lakes. Thus only about 3 million hectares, 7 per cent of the total area, is currently used for farming purposes. Because of the northerly location and mountainous topography, most of the areas suitable for agricultural use are in the southern and central parts of the country. Half of the total arable land is located in the plains along the west and south coasts and surrounding the lakes in central Sweden, and a further almost 40 per cent in the hilly farming and forestry regions in south and central Sweden. Farm holdings in the northern parts of the country consist largely of forests and rough grazing land.

A special feature of the Swedish agricultural sector is the frequent combination of forestry and farming in the same enterprises. Approximately 75 per cent of the Swedish farm holdings (conventionally defined as those containing more than two hectares of arable land) include also some forest land. These holdings account in total for about 25 per cent (5.5m hectares) of Sweden's total forests. Ownership of the other 75 per cent, which is usually not considered as part of the agricultural sector, is divided almost equally between the State or other public entities, companies, and individuals whose holdings do not include two hectares of arable land. The forest land included in farm holdings is on the average more productive, however, since a larger proportion of it is located in southern and central Sweden where annual growth rates are significantly faster. In addition to the combined forestry – farming enterprises, non-farm forestry provides an important source of off-farm employment for part-time Swedish farmers, particularly those in the more remote areas.

Swedish farms, which in 1971 totalled 150,000, are mostly family enterprises, although some food processing industries produce raw materials on their own farms. Almost two-thirds of the arable land is owned by the farmer who uses it, although this proportion had declined as an increasing number of farmers have leased, rather than bought, additional land to enlarge their scale of operations. In 1971 more than 60 per cent of the Swedish farm holders were more than 50 years of age. Farmers and members of their families provide almost 90 per cent of the total labour used on Swedish farms; the number of permanently employed farm workers is estimated to have been reduced to only about 17,000.

Almost half of the Swedish farmers work off their farms for varying periods during the year; in 1971, 15 per cent of them for 1–50 days, 10 per cent for 51–150 days and 22 per cent for more than 150 days. The farmers working off their farms for more than 150 days included roughly 40 per cent of those whose farm holdings included less than 10 hectares of arable land and those whose farm operations did not include cattle, and roughly a third of those less than 50 years of age and those living in northern Sweden.

Sweden's farm production is based largely on livestock. It is estimated that about three-quarters of the crop production is fed to livestock. In 1971, hay and pasture occupied more than 35 per cent of Sweden's total farm land, with feedgrains (barley, oats and mixed grains) accounting for an additional 40 per cent. Most of Sweden's cattle are in relatively small herds. In 1971, 70 per cent of the herds had less than 10 cows and these herds accounted for almost 35 per cent of the total cattle. Pigs tend to be in relatively larger herds, with producers of more than 500 pigs per year accounting for more than half of Sweden's total pig production. Large-scale, specialized production units account for almost all of Sweden's broiler production and are taking over more and more of the egg production.

An important feature of the setting for agricultural adjustment in Sweden has been the strong position and prominent role of the two national farmers' organizations (National Farmers' Union and Federation of Swedish Farmers' Associations), which merged in 1971 to form the Federation of Swedish Farmers. The two organizations, to which almost all Swedish farmers belonged, have represented farmers in negotiations with government authorities concerning prices for farm products. In addition, the farmers' cooperative marketing associations, estimated to handle more than 80 per cent of Swedish farm produce, have played an important role through their centralized buying and selling policies and operations in the implementation of price policies and measures adopted by the government.

The national economic and social policies that have evolved in Sweden since the end of the Second World War have constituted a further major feature of the setting for its agricultural adjustment. Without these policies, the agricultural adjustments that have taken place could hardly have happened, at least not in a manner at all resembling the way in which they did. The aims of these national policies have been full employment, rapid economic growth, and more uniform distribution of income in a balanced economy. Attainment of these aims has been envisaged as involving not merely creation of additional jobs to assure full employment but also redistribution of the Swedish labour force to fill more productive jobs. Since at least the mid-1960s Swedish agricultural programmes and measures have been an integrated part of the government's economic and social programmes, designed to attain these aims withough explicit distinction between the farm and the non-farm sectors. Throughout the period Swedish farmers have been eligible for old-age and disability pensions on the same basis as the rest of the population, and these have enabled the retirement of farmers upon reaching the pensionable age of 67 years or upon becoming disabled.

The rate of Sweden's economic growth has generally maintained a strong demand for labour. With Sweden's low rate of population growth, its labour force has increased by less than one per cent per year, despite a net inflow of foreign labour. The strong demand for labour and consequent opportunities for attractive alternative employment has undoubtedly been the major factor in the rapid transfer of labour to employment outside of agriculture, particularly that of hired farm workers during the 1950s and of children of farm families throughout the period. As the need for additional labour continued during the 1960s, the government adopted an active general labour market policy, with programmes and measures designed to stimulate the movement of labour to more productive employment in dynamic and expanding segments of the industrial and service sectors. The farm population has been fully eligible to participate in these labour market programmes, which have consisted of a wide array of services and benefits designed to increase the mobility of Swedish labour generally.

# 10 Major Agricultural Policy Goals

Guidelines adopted by the Swedish Parliament in 1947 and revised in 1967 have provided the policy framework for Sweden's agricultural adjustment experience since the end of the Second World War. On both occasions a Royal Commission had been appointed previously to study Sweden's agricultural situation and problems, and to make comprehensive policy recommendations. The reports of these commissions generated widespread public discussion and served as the basis for formulation of the guidelines that were subsequently adopted by the parliament. As evidenced by the successive guidelines, Sweden's agricultural policies have evolved to reflect both the increasing industrialization of the economy and urbanization of the population and the continuing improvement of production and management technology on Swedish farms. This evolution has also reflected the continuing national concern for social justice and, during recent years, for conservation of the environment.

Under the 1947 guidelines[1] the priority goal was to provide farm families with the same opportunities as other groups of the population, to obtain a reasonable level of income, and to participate in future increases in the general prosperity. The guidelines envisaged that domestic market prices for major farm products should be stabilized, by means of price and related foreign trade measures, at levels that would yield the desired income parity for farm families.

During the later 1950s it became evident that the Swedish situation was developing somewhat differently from what had been generally anticipated in 1947. The strong demand for labour, which expansion of non-agricultural sectors was creating, had not been fully foreseen; nor had the impact of mechanization and other improved technology, in terms of both agricultural production and the resource requirements of the sector. Production of some farm products exceeded domestic requirements and export of the surpluses had to be subsidized. It had not proved feasible, in practice, to maintain the income parity between farm families and those in other sectors of the economy, as had been envisaged in the 1947 guidelines.

The 1967 guidelines, in contrast to those of 1947 which had tended to consider the agricultural sector as a somewhat autonomous entity, accepted that agriculture should be considered as an integral part of the national economy. Consequently the revised guidelines represented application to the sector of Sweden's national economic and social policies and programmes. Thus, the goals with respect to the income of farm families and the economic efficiency of the agricultural sector were those derived from the national policy objectives of full employment, rapid economic growth and more uniform distribution of income in a balanced economy. In the context also of Sweden's national objectives, these guidelines formulated explicitly a long-term production goal for Swedish agriculture in terms of a self-sufficiency ratio (SSR) for food, based on the need to maintain domestic agricultural production capacity adequate to ensure fulfilment of the country's food requirements in time of emergency.

The 1967 guidelines foresaw the need for re-examination of certain aspects of Sweden's agricultural policies after a period of years, and a third Royal Commission was appointed in 1972 for this purpose. The terms of reference for this commission are more limited than were those for the previous ones, since they anticipate that the policy goals contained in the 1967 guidelines will continue generally unchanged. Although the importance of continued rapid rationalization of Swedish agriculture is reasserted, the commission is not to deal with this subject, except as it may arise in connection with other questions, as new rules concerning these programmes had been adopted by the parliament. Similarly, the commission is not to deal particularly with the question of maintaining agricultural production in the northern parts of the country; this largely involves farm rationalization activities and special price supplements (which are already subject to review every three years). In addition, the commission is not to concern itself directly with conservation of the environment or with the quality of foodstuffs, since other Swedish government agencies have primary responsibility for these matters. Thus, the third Royal Commission is to examine questions relating mainly to the SSR-based production goal and to methods for providing support to Swedish agriculture.

## THE PRODUCTION GOAL

Because of its traditional position of neutrality, the overriding consideration in Sweden's agricultural production goals has been the need to maintain a minimum domestic production capacity adequate to assure fulfilment of the nation's food requirements in time of emergency. The SSR which has been involved in Sweden's agricultural policies is conceptually the ratio between the production capacity (as measured by current production normalized with respect to yields) and Sweden's current consumption requirements, both measured in terms of calories. It is to be noted that this concept of an SSR differs from the usual one which relates actual production during a given period to actual consumption during the same period. Use of the same term for two distinctly different measurements may have led to some misunderstanding in other countries of the Swedish policy goal.

Throughout the period since the end of the Second World War, Sweden's agricultural production capacity has been in excess of the minimum considered to be needed for security purposes. Thus, there has actually been no occasion to invoke the security-based SSR to avoid Sweden's agricultural production capacity being reduced below this minimum level. This does not mean, however, that adoption of the SSR as a policy goal has had no influence. It has, first, served to lessen political opposition to programmes designed to contract the agricultural sector, because it provided assurance that, despite such contraction, Sweden's food requirements in time of emergency could nevertheless be covered. Secondly, it has provided a clear sense of direction and, although less precisely, of the extent of the change desired in the country's food production capacity.

From a practical point of view, Sweden's production goals throughout this period have been primarily to avoid, in so far as possible, production of individual commodities in excess of current domestic requirements, since excess supplies would tend to depress domestic market prices, with consequent adverse effects on farm incomes. With Swedish prices for farm products generally higher than world market levels, financial losses have usually been incurred in the export of Swedish surpluses.

In connection with the 1947 guidelines it was generally agreed that, in time of peace, domestic production to cover 90 to 95 per cent of domestic food requirements (in terms of calories) would be a suitable long-term goal for Sweden. With this level of self-sufficiency in peacetime, it was estimated that Sweden would (with storage and with changes in domestic production and consumption that would result from the emergency conditions) be able to fulfil its food requirements, without imports, in time of emergency.[2] Domestic food production at that time was roughly equivalent to domestic requirements in terms of calories. Such a goal, therefore, implied a reduction of perhaps as much as 10 per cent in Sweden's total agricultural production. Because of the international situation then existing, such a decrease was not considered appropriate and the 1947 guidelines did not actually include an SSR production goal.

According to the 1967 guidelines, Swedish farm production should, by the end of the 1970s, be reduced to 80 per cent[3] of the then current peacetime consumption requirements. During the parliamentary debate that preceded adoption of these guidelines, the Minister of Agriculture explained that although Sweden needed to maintain a certain level of domestic production in order to assure its food supplies during time of emergency and that prices for this domestic production would usually need to be higher than world market levels, domestic production should not exceed that actually required for security reasons, in order to minimize the cost of the subsidies involved.

It was calculated that the difference of 20 per cent between the SSR in peacetime (80 per cent) and that in time of emergency (100 per cent) could be covered as follows:

| | *Percentage in terms of calories* |
|---|---|
| Loss of food imports | −3 |
| Loss of feed imports | −3 |
| Loss of fertilizers imports | −5 |
| Shift in production from animal products to crop products which would more than offset lower yields | +11 |
| Shift in consumption from animal products to crop products | +10 |
| Reduction in per caput consumption | +10 |
| Total | +20 |

According to the terms of reference for the Royal Commission appointed in 1972, the SSR contained in the 1967 guidelines has proved, for a number of reasons, to be an unreliable measure in peacetime of Sweden's agricultural production capacity and capability to fulfil its domestic requirements in time of emergency. A special working group has, therefore, been established to review, and possibly to revise, the methodology used to arrive at the 1967 estimate. On the basis of the working group's report, the commission is to consider whether, because of recent changes in trends and production techniques or for other reasons, the long-term production goal contained in the 1967 guidelines should be revised. The commission is also to investigate methods to be used to avoid regular production in excess of domestic requirements of individual commodities, particularly of cereals.

## THE INCOME GOALS

Under the 1947 guidelines, farm families were to be given the same opportunities as other groups of the population to obtain a reasonable level of income and to participate in future increase in the general prosperity. This did not, however, mean that over-all income parity for Swedish agriculture was to be guaranteed by the government. Rather, it was implied that the desired income[4] parity would be achieved when, in a normal year, the average income of 'basic farms' (well-located, efficiently managed farms with between 10 and 20 hectares of arable land) was equal to that of specified groups of industrial workers. The primary means for attaining the envisaged equality of incomes was stabilization of domestic prices for major farm products at levels higher than would otherwise prevail. Once attained, it was anticipated that this equality should be maintained by periodical reviews, with more or less automatic upward adjustments, of prices for farm products to offset the effects of increases in industrial wages, in domestic price levels, in farm production costs, etc.

This definition of intended income parity was accepted explicitly in 1955, at which time it was considered to have been approximately achieved. It did not, however, prove possible for it to be maintained in subsequent years. As the income of industrial workers continued to rise and farm production costs to increase it was not considered feasible to allow fully compensating increases in the prices for farm products, and increases in agricultural production and productivity were not sufficient to offset the shortfall. In addition, prices for farm commodities on the world market trended downward in relation to those in the protected home market, increasing the losses incurred on exports of Swedish surpluses. Since these losses were to a large extent covered by production fees paid directly by farmers, this development imposed a further restraint on farm income.

The 1967 guidelines retained the general aim of providing opportunities for farmers and their families to attain the same level of economic and social well-being as those engaged in other sectors of the economy. Whereas the earlier guidelines had envisaged exact comparisons and equality between the average income of certain categories of farmers and of specified groups of industrial workers, income goals in the 1967 guidelines were expressed as 'reasonable' incomes for those employed in agriculture. Moreover, whereas the income goals envisaged in the 1947 guidelines related only to certain categories of farmers (initially 'basic farms' and later 'norm farms')[5] those in the revised guidelines contained no limitation with respect to the size of farm. Although not explicitly stated, the income goals now tended to relate mainly to families operating larger-scale, rationalized farms, with the income position of families with small, non-viable holdings considered to be primarily the concern of national economic, social and welfare programmes, rather than of agricultural policies.

Under the revised guidelines, the rationalization of farm structures and operations (rather than price and related foreign trade measures) was the principal means for attaining the more loosely formulated income goals for the reduced population that would continue to be engaged in agriculture. The income goal was further modified in July 1971 for farmers in northern Sweden and a year later for dairy farms in central and southern Sweden who, although their holdings did not meet the usual criteria for economic viability, could expect, with a limited type of rationalization support, to obtain a 'decent income', at least half of which would be from the farm holdings.

As noted above, the Royal Commission appointed in 1972 is to study and make recommendations concerning alternative methods for continuing to support Swedish agriculture. With respect to price support, the 1967 guidelines directed that in principle a so-called 'high price line'[6] was to apply but emphasized that possible change to a so-called 'low price line'[7] should be re-examined after a period of time. Thus, the commission is to investigate the implications of such a change, including the possibility of shifting to some combination of the two (a 'middle price line'). The alternative methods are to be examined for their likely effects on, among other things, the income of farm families (especially of those operating larger-scale, rationalized farms who account for the greater part of Sweden's total farm production), on consumers (particularly those in lower-income brackets and families with children who spend a larger part of their disposable income on food), on the food processing industries, on government finances (deficiency payments with the low price line could constitute a relatively heavy additional burden on the budget), on the continued rationalization of agriculture, on the production of individual farm commodities, and so on. The commission is also to consider the possibility of increasing investment support (in the form of capital grants and other investment subsidies) in order that price support might be limited.

## THE ECONOMIC EFFICIENCY GOALS

The 1947 guidelines, at least implicitly, sought increased economic efficiency within the agricultural sector, in terms of both the operation of individual farms and the productivity of resources in the sector. Such increased efficiency was to result mainly from the rationalization of farm structures and operations. At the farm level, the envisaged rationalization included enlargement of individual farms and improvement in their layout (external rationalization), permanent improvement of the agricultural land and farm buildings (internal rationalization) and increases in the efficiency and profitability of individual farm operations (rationalization of farm management). At the sector level, labour and marginal land were to be withdrawn from agricultural production. Small non-viable farms were to be discontinued in all districts, although production on economically efficient farms was to be maintained in as much of the country as possible.

The process of farm rationalization came increasingly to be considered as moving too slowly. Certain economists and politicians contended that large benefits for the community could be obtained by a more rapid transfer of production factors, especially of labour, from agriculture to other more productive sectors. These arguments were, in fact, based on the growing shortage of labour for these other sectors rather than on the availability of lower-priced imports of agricultural products. In any case, rapid progress in mechanization and other improved technologies appeared to give additional advantages to larger-scale enterprises in both agriculture and forestry.

On the other hand, the rapid movement of farm families out of agriculture was creating economic, social and political problems, particularly in more remote areas including northern Sweden. Farms in the forest and mixed farm – forest areas were being closed down at a rapid rate, which was expected to accelerate because of the high average age of farmers. It was becoming evident that rational utilization of abandoned arable and forest land would become increasingly a problem. The depopulation and economic stagnation of rural areas was making it increasingly difficult and expensive to maintain public services (schools, medical care, transportation, communications, etc.) and the social and commercial infrastructure. In 1960 special rationalization programmes (SR-activity) were introduced in northern Sweden to deal with these and related special problems.

The Land Acquisition Act of 1965 represented an important change in Sweden's agricultural land policies. In contrast to previous legislation, this Act contained no absolute obstacles to acquisition of agricultural land by other than active farmers.[8] Rather, it indicated conditions under which County Agricultural Board (CAB) approval for acquisition of agricultural land might be refused, after examination of the relevant circumstances on a case-by-case basis, and provided some general priorities which the CAB were to apply in reaching their decisions.

However, under the guidelines adopted in 1967, the creation of rationalized family holdings has in practice continued to form the main object of the rationalization activities, and a sizable agricultural production has been maintained in northern Sweden. The programme adopted in 1967 aimed, however, at the removal of restrictions which in the past had tended to hamper the formation of sufficiently large farm holdings. All efficient agricultural enterprise (regardless of size or type, or category of owner) were made eligible to receive government support for farm enlargement and other rationalization purposes.

NOTES

1 The 1947 guidelines were based on the report made in 1946 by the Royal Commission that had been appointed in 1942. An earlier Royal Commission with similar terms of reference had been appointed in 1938 but had been disbanded when its work was disrupted by the outbreak of the Second World War.

2 In this context, an emergency was defined as a period of up to three years during which Sweden would, for any reason, be cut off from imported supplies.

3 Although Swedish sources usually refer to the SSR as a specific percentage, the procedures involved in arriving at this ratio make it appropriate to treat it as a range covering roughly 5 percentage points. The apparent reduction of 5–10 per cent, as compared with the SSR generally agreed but not formally adopted in 1947, seems to have been mostly a result of modifications in the procedures used to estimate this ratio.

4 In this context, 'farm income' was interpreted as consisting of the labour income of the operator and his family from farm and non-farm sources plus certain parts of their income from capital.

5 Well-located, efficiently managed farms with between 20 and 30 hectares of arable land.

6 With a high price line, domestic producer prices are separated from world market prices and supported at higher levels, with the cost of the support passed on to consumers as part of the retail price.

7 With a low price line, domestic producer prices are linked to world market prices, and support to domestic producers consists of some sort of deficiency payments from the government.

8 A primary aim of Sweden's agricultural land policies was, until 1965, to preserve arable land and farm-associated forestry in the hands of active farmers and to maintain the family farm as the basic and desirable unit for agricultural production. The Land Acquisition Acts of 1945 and 1948 provided regulations to prevent acquisitions of agricultural land by private persons for investment, or speculation and otherwise undesirable purposes. The Land Acquisition Act of 1955 required that permission be obtained from the County Agricultural Board for all acquisitions of farm real estate by private persons (except in cases of inheritance, legacy or a buyer being closely related to the seller).

# 11 Résumé of the Programmes and Measures

The evolution of Sweden's agricultural goals has been accompanied by generally comparable modifications in its agricultural programmes and measures. Throughout the period under review, programmes aimed directly at attaining the income goals have consisted mainly of price and related foreign trade measures, and those aimed at accomplishing the economic efficiency goals have consisted of measures for the rationalization of farm structures and operations. There have been no programmes or measures aimed directly at attainment of the SSR production goals. Swedish agricultural programmes have, moreover, included special measures designed mostly to benefit small farmers and especially those in the northern areas.

## PRICE AND RELATED FOREIGN TRADE MEASURES

Throughout the period under review, price and related foreign trade measures have served to stabilize domestic market prices for Sweden's principal farm products at predetermined levels, generally above those on world markets. Series showing comparative trends in producer prices, consumer prices and world market prices for commodities since 1966/67 are contained in Table 11.1.[1] Additional measures, including export subsidies, have been required when domestic production of a specific commodity has exceeded domestic requirements. The National Agricultural Marketing Board, functioning under the direction of the Ministry of Agriculture,[2] has been in charge of the government's farm price regulation activities throughout the period under review. A characteristic feature of these activities has, however, been the delegation of authority and decentralization of administration to the farmers' cooperative marketing associations and the special Marketing Associations (also non-governmental bodies) that have been created for each of the major farm commodities.

Until 1 September 1956, fixed prices were guaranteed by the government for Sweden's domestic farm products. These prices were negotiated each spring between the farmers' organizations and the government authorities, with a consumer delegation advisory to the latter. The price

TABLE 11.1 Indices of prices for farm products

| | 1967/68 | 1968/69 | 1969/70 (1966/67 = 100) | 1970/71 | 1971/72 |
|---|---|---|---|---|---|
| *Producer price index* | | | | | |
| Crop products | 100 | 95 | 104 | 90 | 94 |
| Livestock products | 101 | 101 | 104 | 108 | 120 |
| All products | 101 | 100 | 104 | 105 | 115 |
| *Price regulation index*[a] | | | | | |
| Crop products | 95 | 96 | 111 | 105 | 114 |
| Livestock products | 100 | 101 | 106 | 109 | 119 |
| All products | 99 | 100 | 107 | 108 | 118 |
| *World market price index* | | | | | |
| Crop products | 90 | 91 | 95 | 114 | 117 |
| Livestock products | 91 | 92 | 103 | 103 | 119 |
| All products | 91 | 92 | 102 | 104 | 119 |
| *Index of prices for inputs* | | | | | |
| Requisites | 103 | 105 | 106 | 108 | 110 |
| Services | 105 | 105 | 108 | 115 | 123 |
| Capital costs | 103 | 102 | 107 | 113 | 117 |
| Total | 103 | 103 | 107 | 112 | 115 |
| *Consumer price index* | | | | | |
| Food from price-regulated farm products | 102 | 104 | 110 | 120 | 133 |
| Other food | 102 | 103 | 111 | 119 | 126 |
| Total[b] | 104 | 106 | 111 | 119 | 126 |

[a] Price at the stage where the agricultural pricing regulations are effective.
[b] Excluding food from price-regulated farm products.

negotiations were based on a total budget for the 'national farm', i.e. for the aggregate of all the farm holdings in the agricultural sector. Income and costs during the coming production season were estimated, assuming average weather conditions and stable prices, wages and interest. In the autumn the budget was recalculated, taking into account the actual harvest results. If the revised total agricultural revenue differed from that estimated in the spring by more than 4 per cent, commodity prices (and/or subsidies) were adjusted to reduce the surplus or deficit to a maximum of 4 per cent. During this period frontier protection for Swedish agriculture was provided by an array of custom duties, variable import levies, import monopolies, quantitative controls, etc. The variable import duties were adjusted to keep the cost of imported supplies in line with the predetermined level of domestic market prices. The quantitative import controls were exercised through import monopolies granted to the Marketing Associations or through State trading operations.

Effective from 1 September 1956, the fixed prices were largely replaced by more flexible support prices with the levels continuing to be based on negotiations. In addition to the total budget for the 'national farm', the profitability of 'basic farms' ('norm farms' after 1965) were also taken into account. This procedure resulted in long-term Agricultural Agreements, ratified by the parliament; for three years (1956–9) in the first instance, for six years (1959–65) in the second.[3] At the beginning of each Agricultural Agreement period, 'middle prices' (wholesale prices relating to closely specified market quotations) were fixed for each of the main farm products, with safeguard provisions designed to assure that these 'middle prices' would be adjusted upward as industrial wages rose, farm production costs increased, etc. During the period of the agreement, domestic market prices were allowed to fluctuate within a range extending, according to the product, 10 to 15 per cent on either side of the 'middle price'. For breadgrains (wheat and rye), oilseeds and industrial potatoes, supplementary market stabilization measures were provided to avoid a possible sharp drop in prices during seasonal supply peaks.

Also effective from 1 September 1956, import monopolies and quantitative import controls were, in principle, terminated and the variable import levies replaced by fixed import levies.[4] The fixed import levy was determined as the difference between the 'middle price' established for the commodity and the comparable world market price that prevailed at the beginning of the Agreement period. So long as fluctuations in the domestic prices remained within the range that had been established with reference to the 'middle price', this levy remained unchanged. If, however, the domestic price moved outside this range, the fixed levy was to be adjusted and, if this proved not to be sufficient, the reintroduction of import controls or any other measures considered appropriate should be taken to bring it back inside the prescribed range.

Under the guidelines adopted in 1967, price and related foreign trade measures of the same general nature as those included in the 1959 Agricultural Agreement continued to be applied. The 1967 guidelines, however, provided greater flexibility in determination of the levels at which domestic market prices were to be stabilized. These levels continued to be based on negotiations between the government authorities (with an advisory consumer delegation) and the farmers' organizations, which take into consideration, among other things, the market situation, the farm income position, the progress of farm rationalization and of productivity in the agricultural sector, etc. It was considered necessary, in the short run, to restrain increases in farm prices in order to facilitate and to expedite the desired adjustment of production, the rationalization of farm structures and operations and the transfer of resources from agriculture to other uses. Import levies continued to provide protection at the frontier for most of Sweden's major farm products. The automatic rules regulating the amount of these levies were, however, largely abolished.

Under the 1967 guidelines, price regulations for major Swedish farm products were adopted covering successively the two-year periods 1967–9 and 1969–71, and the three-year period 1971–4. The last of the regulations provided for automatic increases in the 'middle prices' for the second and the third years, with additional increases possibly to be granted each 1 January to compensate for increases in the Swedish general price level. To curb the inflationary spiral, consumer prices were frozen from 1 January 1973 and the cost of the increase in producer prices which became effective at that time is being borne directly by the government budget, rather than indirectly by consumers through higher retail prices.

### FARM RATIONALIZATION MEASURES

Rationalization of farm structures and operations have been the principal direct means for improving the efficiency of the Swedish agricultural sector. It has been anticipated that this rationalization would, at least in the long run, result in the transfer of resources (and especially labour) out of farming, a reduction in the total production capacity of Sweden's farms (as compared with capacity in the absence of these measures), and improved opportunities for those families who continued in farming to obtain reasonable levels of income.

Under the Ministry of Agriculture, the National Agricultural Board has had responsibility for the direction of official efforts in the areas of rationalization, planning, vocational training and advisory services, and has supervised and coordinated the activities for this purpose of the twenty-four County Agricultural Boards. The CABs have administered the financial assistance that has been available from the government for external and internal rationalization of individual holdings. They have engaged in active land purchase operations and have ensured compliance with the provisions of the series of land acquisition laws. They have also developed general plans for the rationalization of farm structures within their respective counties and revised such plans to reflect changed conditions (technological, economic, social, etc.). In addition, they have performed extensive advisory work in support of the rationalization activities. Swedish government financial assistance for the rationalization of farm structures and operations is summarized in Table 11.2.

Financial assistance from the government has been available to owners of individual holdings for both external and internal rationalization measures. This assistance has been in the form of grants and of guarantees for the repayment of loans obtained from usual credit sources. The guarantees have enabled the borrowers usually to obtain more favourable interest rates and repayment terms from these usual credit sources. They have also frequently been able to obtain larger loans for the purchase of land. Loans without government guarantees were normally based on the value of the land as assessed for tax purposes, while loans with government guarantees tended to reflect the current market value of the land; the sales value of the agriculrural land sold in 1971 exceeded the assessed value by roughly 70 per cent. Grants and guarantees combined could cover up to 100 per cent of the costs of enlargement or of the internal rationalization improvements.

The CAB's active land purchase operations have consisted of the actual purchase of farm land and forests and the resale of this land for enlargement, or improvement in the layout, of individual holdings. These land purchase operations have been financed by the Land Fund, established in 1948. In principle, the land was to be resold as soon as possible to farmers who needed it. It is estimated that about 40 per cent of the land which changes hands actually passes through CAB ownership, although in northern Sweden it is estimated as about 90 per cent. Arable land in southern Sweden, particularly in the flatlands, tends mostly to be bought directly by neighbouring farmers with the CAB role in the transactions largely limited to that of guaranteeing the loans involved.

Under the 1947 guidelines, the rationalization activities of the CABs were initially directed largely to encourage farmers to undertake voluntary rationalization and to help them to achieve effective results. The Boards acted frequently as intermediaries for direct transactions between farmers. Their activities were aimed mainly at the enlargement of 'transitional farms' (mostly with 2 to 10 hectares of arable land) up to the size of 'basic farms' (those with 10 to 20 hectares of arable land).

Under the Land Acquisition Act of 1955, the conditions governing the CABs' direct intervention in the agricultural land market were substantially modified. Their active land purchase operations expanded significantly and, in 1959, became directed towards the creation of 'norm farms' and of 'two full-time worker farms'. The government rationalization grants and guarantees were also made available for this purpose.

TABLE 11.2 Government financial assistance for farm rationalization

| | Type of assistance | | | Purpose of assistance | | | | |
|---|---|---|---|---|---|---|---|---|
| | *Total* | *Loan guarantees* | *Grants* | *Internal rationalization*[a] | *External rationalization*[b] | *Land acquisition loans*[c] | *Operating loans* | *Machinery loans* |
| | | | | (million Kr.) | | | | |
| 1950 | 58 | 43 | 15 | 24 | 5 | 22 | 7 | — |
| 1951 | 52 | 39 | 13 | 20 | 5 | 21 | 6 | — |
| 1952 | 54 | 38 | 16 | 23 | 5 | 20 | 5 | — |
| 1953 | 72 | 54 | 18 | 28 | 7 | 29 | 8 | — |
| 1954 | 70 | 52 | 18 | 29 | 9 | 25 | 7 | — |
| 1955 | 59 | 43 | 16 | 27 | 8 | 18 | 5 | — |
| 1956 | 59 | 44 | 15 | 24 | 9 | 20 | 6 | — |
| 1957 | 65 | 50 | 16 | 26 | 10 | 23 | 6 | — |
| 1958 | 54 | 40 | 14 | 21 | 9 | 19 | 5 | — |
| 1959 | 58 | 43 | 15 | 22 | 11 | 18 | 6 | — |
| 1956 | 61 | 49 | 12 | 18 | 14 | 21 | 7 | — |
| 1961 | 70 | 56 | 14 | 21 | 17 | 24 | 8 | — |
| 1962 | 81 | 67 | 14 | 23 | 24 | 25 | 9 | — |
| 1963 | 93 | 78 | 15 | 28 | 32 | 23 | 10 | — |
| 1964 | 115 | 95 | 20 | 41 | 34 | 27 | 13 | — |
| 1965 | 134 | 111 | 23 | 52 | 42 | 26 | 15 | — |
| 1966 | 164 | 141 | 23 | 61 | 57 | 28 | 18 | — |
| 1967 | 181 | 158 | 23 | 61 | 63 | 35 | 21 | 1[d] |
| 1968 | 222 | 199 | 23 | 78 | 69 | 35 | 38 | 3 |
| 1969 | 181 | 162 | 19 | 68 | 55 | 22 | 34 | 2 |
| 1970 | 151 | 131 | 19 | 55 | 44 | 21 | 29 | 1 |
| 1971 | 173 | 148 | 25 | 67 | 44 | 22 | 39 | 2 |
| 1972 | 220 | 186 | 34 | 100 | 51 | 22 | 45 | 1 |

[a] For ditching, draining and embanking, under-draining, fixed equipment, and farm buildings.
[b] For acquisition of additional land to enlarge present holdings. [c] For acquisition of farm holdings. [d] Six months only.

The internal rationalization grants, however, became subject to the requirement that the proposed improvements should be economically justified and limited to farms which in the future would be economically viable production units.

The change in Swedish policy concerning the ownership of agricultural land, contained in the Land Acquisition Act of 1965, had important implications for the nature and scope of the CABs' rationalization activities. Although the law gave a first priority in the acquisition of agricultural land to active farmers, a relatively wide margin was left for the exercise of discretion by the CABs. Under these guidelines, the CABs' rationalization activities were no longer directed towards creating farms of any particular size, but rather towards the creation of efficient agricultural enterprises, regardless of size, type, or category of owner. The use of grants was restricted, and the normal form of government financial assistance for rationalization became the guaranteeing of repayment of loans obtained from usual credit sources.

Special rationalization programmes (usually referred to as SR-activity) were introduced in 1960 in the four most northern counties and later extended to four more counties in northern Sweden. Because of the local economic situation and the social and demographic conditions, it had been decided that special efforts were required in these counties. Depopulation and economic stagnation were common characteristics for large areas in the SR counties. The agricultural structure was dominated by non-viable units in great need of comprehensive investments in soil improvement, buildings, livestock, machinery, etc. The total investment needed was such that the conventional types of government support were insufficient to meet the requirements.

The SR-activity has been planned and carried out at the county level by the CAB in close collaboration with the Rural Land Survey Council and the Forestry Council. The aim has been to establish viable farm units as rapidly as possible by simultaneous application of structural, technological and management rationalization measures combined with extensive advisory services. The goal (in 1970) has in most cases been a combined enterprise of some 60 to 70 hectares of arable land and 400 to 500 hectares of forest land. Investments of the order of 0.6 to 1 million Swedish Kr. per farm (equivalent to roughly US $100,000–200,000 at the 1970 exchange rate) have been involved in the building up of such units, equipped to exploit modern labour-saving techniques.

The rate of SR-activity tended to decline during the later 1960s because of rising interest rates, credit restrictions, increasing cost for buildings and labour, etc. A government committee, reporting in 1970 on agricultural conditions in northern Sweden, identified the major obstacles to expansion of SR-activity as the difficulties in reaching satisfactory profitability positions and the risk of liquidity crises for SR-farms. It also noted difficulties in some areas in finding suitable farmers willing to assume financial obligations of the magnitude involved in creating an SR-farm.

Since July 1971 a limited type of rationalization support (B-support) had been available to a special category of farmers. They are usually 45 to 50 years of age, with holdings that are not currently needed for enlargement of other farms. Although their holdings do not meet the usual criteria for economic viability, the operator can, with the limited rationalization support, expect to obtain a 'decent income', at least half of which must come from the farm holding. The B-support was made available initially in the northern areas to halt disruption of the infrastructure as a result of the rapid depopulation. Since July 1972 similar support has been available to milk-producing farmers (with a minimum of 500,000 kg per year) in southern and middle Sweden.

## SPECIAL MEASURES

In addition to the above measures which constitute the main effort of Sweden's agricultural programmes, a number of special measures have also been applied, mostly for the benefit of small farmers and especially of those in the northern areas, and to promote movement of farm families out of agriculture.

An acreage subsidy, scaled to the size of their holdings or arable land and subject to various conditions, was paid to supplement the incomes of operators of small farms (2 to 10 hectares of arable land). The maximum payment was Kr. 500 per year, and its importance as a source of supplementary income was therefore relatively limited. In 1967 the acreage subsidy payments were replaced by the 'transition support' scheme (see below).

A delivery supplement, scaled to the volume of milk delivered to creameries, was paid to provide a higher return for milk delivered by small-scale producers. At national average production per cow, the highest payments were received by farmers with 3 to 5 cows and those with more than 8 cows received none. The maximum payment under this scheme was Kr. 400 per year and its importance as a source of supplementary income was also relatively limited. These supplementary delivery payments were also replaced by the 'transition support' scheme in 1967.

A special milk price supplement has been paid to farmers in the northern areas, to compensate for higher production costs and for the higher collection and creamery operation expenses. This supplement has represented about 18 per cent of the normal price and applied to the total volume of milk delivered. This scheme has been considered mainly as a form of regional support and was continued in effect after 1967. Similar price supplements for beef and pigmeat were introduced in July 1971.

In 1967 the acreage subsidies and the supplemental milk delivery payments were replaced by a 'transition support' scheme, designed to facilitate the transfer of elderly farmers with smallholdings to other employment or early retirement. To qualify for transition support, farmers were required to have been receiving at least one of the replaced payments, to have started the farm operation before 1967, to have 2 to 20 hectares of arable land, and not to be older than 67 years (the pensionable age). In 1969 annual payments under this scheme averaged Kr. 550 per person. Persons eligible for this support are under certain circumstances entitled to exchange the annual support for a lump-sum retirement compensation, scaled according to the age of the recipient. The scheme of transition support was to be phased out gradually, with no support after 1970 for persons born after 1912.

In addition to the transition support, farmers wishing to leave agriculture have also been eligible for assistance and services available under Sweden's general labour market policy. These have included vocational guidance and information concerning alternative employment opportunities, retraining courses with allowances paid during the period of training, transfer grants to assist farmers and their families to move to the location of new employment, regional development assistance and grants payable to employers for the purpose of creating additional employment opportunities, and so on. These activities have been administered by the County Labour Boards in coordination with the CABs.

## NOTES

1 Although the level of producer prices has risen, the rate of increase has since 1966/67 been notably less than that in consumer food prices and in Swedish price indices generally. Thus, in 'real terms', producer prices have in fact declined, at least since 1966/67.

2 See Annex, p. 201, for a description of the National Agricultural Marketing Board, and other organizations mentioned throughout this study.

3 The six-year contract was extended subsequently for two additional periods of one year each, to make a total of eight years (1959–67).

4 However, the system of fixed import levies did not include oilcake, maize and certain other feedstuffs, for which variable import levies continued to be applied.

# 12 Adjustments that have Taken Place

During the period under review, adjustments have brought important changes in Sweden's farms, in its agricultural sector, in the position of this sector in the national economy, and in Sweden's foreign trade in agricultural products. At the farm level, adjustments have consisted of changes in the dimensions, structure, equipment, organization and operations of individual farms. The net effect of these has been to alter the dimensions of the agricultural sector in terms of both resources and production, and the structure of the sector especially in terms of the number and size of farms. At the national level, the totality of adjustments on individual farms has diminished the relative position of the agricultural sector in the Swedish economy and increased its linkages with the rest of the economy. Adjustments at the international level have been largely the external trade counterparts of changes that have taken place in the available supplies of domestically produced farm products.

## AT THE FARM LEVEL

Adjustments at the farm level have constituted the fundamental element in the Swedish experience. The primary motivation for adjustments at this level is of course the farmer's desire to improve the profitability of his farm and thereby to increase his income. The substantial enlargement of the scale of operations that has taken place has, with fuller utilization and higher productivity of the resources available, increased this profitability. These adjustments have moreover contributed also to increased economic efficiency of resource utilization in the agricultural sector. Since Sweden concurrently sought to avoid increasing its agricultural production, the number of farms has had to be reduced to allow for enlargement of those that remained; and the aggregate of agricultural resources also reduced to accommodate the higher productivity.

The adjustments at the farm level have, as might be expected, differed widely as between individual farms. A relatively large number of Swedish farmers have chosen to leave agriculture for employment in other sectors, or for retirement. The number of farm holdings dropped from 282,000 in 1951 to 150,000 in 1971, i.e. by almost half in twenty years. Although it is not possible to determine exactly the characteristics of those who left agriculture, it appears that a large proportion of them were from holdings that included relatively small areas of arable land, were in the lower income brackets and were in areas less well suited by natural conditions for farming purposes. Between 1960 and 1970 there was an estimated net reduction of 177,000 (43 per cent) in the number of persons 15 years of age and older who were employed in agriculture. Almost half of this net reduction was accounted for by transfers to other employment, and the remainder by withdrawals from the labour force (retirement on pensions 30 per cent, deaths 20 per cent and disabilities 7 per cent).

Adjustments that have taken place at the farm level are reflected in the changes that have occurred in the average dimensions of Swedish farms (Table 12.1). The average size of Swedish farms has increased sub-

TABLE 12.1 Average dimensions of Swedish farms, holdings with more than 2 hectares of arable land

| | *1951* | *1961* (*hectares*) | *1971* |
|---|---|---|---|
| RESOURCES | | | |
| Land, total | 52.1 | — | 68.7 |
| *of which*: arable | 12.5 | 14.2 | 20.1 |
| forest | 28.3 | — | 36.9 |
| | | (*persons*) | |
| Labour, total | 3.1 | 2.3 | 1.5 |
| operators and unpaid family | 2.4 | 1.9 | 1.2 |
| hired | 0.7 | 0.4 | 0.3 |
| | *1953* | *1961* (*Kr.*) | *1971* |
| Capital assets, total | 61,819 | 103,010 | 203,300 |
| real estate | 37,211 | 62,026 | 98,600 |
| non-real estate | 24,608 | 40,984 | 104,700 |
| PRODUCTION | | | |
| Value of products sold | 16,539 | 24,334 | 53,285 |
| crop products | 4112 | 5100 | 14,716 |
| animal products | 10,036 | 15,529 | 32,728 |
| forest products | 2391 | 3705 | 5841 |
| Cash production expenditures | 11,902 | 19,140 | 47,393 |
| NET CASH INCOME | | | |
| Farm sources[a] | 6582 | 8574 | 12,020 |
| Other sources[b] | 1327 | 2988 | 12,150 |
| Total | 7909 | 11,562 | 24,171 |

[a] Includes incomes derived from forest land contained in the farm holding.

[b] Includes income of the operators and members of his family from employment on farms other than his own and in forestry other than that included in his farm holding.

stantially as measured in terms of land area, of capital assets, of the value of products sold, or of the cash production expenses, and has decreased in terms of the labour inputs. Output per person employed in agriculture has therefore risen. The net cash income of farm families has increased, particularly as a result of larger incomes from non-farm sources[1] which had represented only 17 per cent of the total in 1951 but had risen to slightly more than 50 per cent in 1971. Income from non-farm sources accounted for no less than 85 per cent of the 40 per cent increase in the total net cash income of farm families that occurred between 1966 and 1971. The greatly increased importance of non-farm sources of income for families living on farms is one of the most striking features of the changes that have taken place in the Swedish situation during recent years. It underlines the necessity for looking at agricultural policies in a setting broader than that of farming operations alone.

Further insight into the results of the adjustments that have taken place at the farm level is provided by the average dimensions in 1971 of Swedish farms grouped according to the hectares of arable land contained in the holding (Table 12.2). Despite a decrease between 1951 and 1971 of more than 55 per cent in the number of Sweden's farms with less than 20 hectares of arable land, 106,000 Swedish farm holdings (70 per cent of the total) contained less than this area of arable land in 1971. A large part of these have, however, become largely part-time farms; more than 60 per cent of their net cash income was obtained from non-farm sources. They accounted, nevertheless, for almost a third of the Swedish agricultural sector's arable land, more than half of its total labour inputs and about 45 per cent of its total capital assets. Market sales from these largely part-time farms accounted, however, for less than a third of the sector's total and consisted mainly of livestock products. The possibilities for mechanization and other application of improved technology in crop production tends to be limited to larger-scale farms. Relatively small herds continue, however, to account for a rather large part of Sweden's livestock population; in 1971 about 45 per cent of the cattle were in herds with less than 10 cows and almost 40 per cent of the pigs in herds of less than 100. Because of their relatively large net cash income from non-farm sources, the average family income position of the 106,000 largely part-time farmers in 1971 appears to have compared not unfavourably with that in other sectors of the Swedish economy.

### AT THE SECTOR LEVEL

As measured by its production resources, the Swedish agricultural sector has undergone a notable contraction during the period under review (Table 12.3). Between 1951 and 1971 the area of arable land was reduced by 15 per cent and that of farm – forests by almost a third. The number of

TABLE 12.2 Average dimensions of farms grouped according to the hectares of arable land contained in the holding, 1971

| | Hectares of arable land | | | | | | |
|---|---|---|---|---|---|---|---|
| | 2.1–10.0 | 10.1–20.0 | 20.1–30.0 | 30.1–50.0 | 50.1–100.0 | Over 100 | All farms |
| NUMBER OF FARMS | 69,774 | 36,685 | 17,799 | 14,920 | 8132 | 2704 | 150,014 |
| RESOURCES | | | (hectares) | | | | |
| Land, arable | 5.9 | 14.8 | 24.9 | 38.8 | 67.8 | 181.0 | 20.1 |
| Labour[a] | | | (hours) | | | | |
| Operators and family | 1521 | 2881 | 3376 | 3431 | 3276 | 2452 | 2389 |
| Hired | 102 | 208 | 285 | 497 | 1173 | 6289 | 358 |
| Total | 1623 | 3089 | 3661 | 3928 | 4449 | 8741 | 2747 |
| Capital assets[b] | | | ('000 Kr.) | | | | |
| Real estate | 52.7 | 84.3 | 114.9 | 157.1 | 234.3 | 624.4 | 98.6 |
| Non-real estate | 48.7 | 83.5 | 116.1 | 179.9 | 265.7 | 852.8 | 104.8 |
| Total | 101.4 | 167.8 | 231.0 | 337.0 | 500.0 | 1477.2 | 203.4 |
| VALUE OF PRODUCTS SOLD | | | | | | | |
| Crop products | 1.3 | 5.6 | 14.0 | 31.2 | 71.5 | 225.5 | 14.7 |
| Livestock products | 9.3 | 30.5 | 49.4 | 67.3 | 88.9 | 188.3 | 32.7 |
| Forest products | 3.8 | 5.4 | 5.8 | 7.0 | 9.5 | 46.6 | 5.8 |
| Total | 14.4 | 41.5 | 69.2 | 105.5 | 169.9 | 460.4 | 53.3 |
| CASH PRODUCTION EXPENSES | 12.3 | 33.6 | 57.6 | 93.1 | 159.4 | 473.5 | 47.4 |
| NET CASH INCOME | | | | | | | |
| Farm operations | 5.7 | 12.8 | 18.0 | 21.1 | 24.5 | 32.5 | 12.0 |
| Non-farm sources | 14.9 | 9.9 | 8.4 | 7.9 | 10.8 | 28.2 | 12.2 |
| Total | 20.6 | 22.8 | 26.4 | 28.9 | 35.3 | 60.7 | 24.2 |

[a] 1970.
[b] Value of capital assets as assessed for tax purposes; real estate includes that used for both agricultural and other purposes; market value of agricultural real estate sold in 1971 was approximately 70 per cent above the assessed value.

TABLE 12.3 Resources in the agricultural sector[a]

| | 1951 | 1961 | 1971 |
|---|---|---|---|
| | | ('000 hectares) | |
| Land, total | 14.886 | — | 10.310 |
| *of which*: arable | 3.527 | 3.296 | 3.016 |
| pastures | 179 | 162 | 141 |
| forest | 8.147 | — | 5.539 |
| other | 3.033 | — | 1.613 |
| | | ('000 persons)[b] | |
| Labour, total | 869 | 533 | 230[c] |
| *of which*: operators and families | 674 | 456 | — |
| full-time employees | 90 | 40 | — |
| part-time employees | 106 | 37 | — |
| *of which*: men | 505 | 318 | — |
| women | 364 | 215 | — |
| | *1953* | *1961* | *1971* |
| | | (million Kr.) | |
| Capital assets,[d] total | 17,093 | 23,993 | 29,361 |
| *of which*: real estate | 10,289 | 14,447 | 14,234 |
| livestock | 2017 | 2515 | 2776 |
| machinery and equipment | 1371 | 1886 | 2942 |
| stocks | 404 | 508 | 792 |
| securities, claims | 2484 | 3635 | 6590 |
| other | 528 | 1002 | 2027 |

[a] Aggregate of farm holdings with more than 2 hectares of arable land.
[b] Persons 15 years of age and older.
[c] Estimated.
[d] Value of capital assets as assessed for tax purposes; market value of agricultural real estate sold in 1971 was approximately 70 per cent above the assessed value.

persons 15 years of age and older employed in agriculture was reduced by an estimated three-quarters. Although the value of non-real estate capital assets in the sector was increased by about 120 per cent at current prices, the increase in real terms (at constant prices) was certainly of much smaller proportions. The reduction of resources in the sector would have been a means for decreasing total agricultural production, if the production-decreasing effects had not been approximately offset by increased efficiency of resources utilization on the individual farms that remained. The very large increases in crop and livestock yields, which were probably greater than had been anticipated, are shown in Table 12.6. Since the reduction in the number of farms was not accompanied by a reduction in the total net farm income, these adjustments have increased the average cash income per farm family from its farming operations.

As measured by FAO indices[2] Sweden's total agricultural production in 1970–2 averaged 11 per cent above the 1955–7 level but, on a per caput

TABLE 12.4 Production from the agricultural sector

| | *1950–2 average* | *1955–7 average* | *1960–2 average* | *1965–7 average* | *1970–2 average* |
|---|---|---|---|---|---|
| | | | *(1961–5 = 100)* | | |
| FAO PRODUCTION INDICES | | | | | |
| Agricultural, total | 102[a] | 96 | 100 | 100 | 107 |
| per caput | 109[a] | 99 | 101 | 97 | 100 |
| Crops | 82[a] | 83 | 96 | 101 | 124 |
| Livestock products | 99[a] | 96 | 100 | 99 | 93 |
| | | | *('000 m.t.)* | | |
| CROPS | | | | | |
| Wheat | 648 | 784 | 933 | 931 | 1068 |
| Rye[b] | 221 | 216 | 193 | 152 | 311 |
| Barley | 256 | 517 | 881 | 1465 | 2033 |
| Oats | 743 | 807 | 1225 | 1296 | 1820 |
| Mixed grains | 682 | 614 | 519 | 312 | 209 |
| Total cereals | 2550 | 2938 | 3751 | 4156 | 5441 |
| Oil seeds[c] | ... | ... | ... | 193 | 281 |
| Potatoes | 1530 | 1456 | 1514 | 1408 | 1315 |
| Sugar beet | 1769 | 1851 | 1952 | 1524 | 1680 |
| | *1950/51–51/52* | *1955/56–56/57* | *1960/61–61/62* | *1965/66–66/67* | *1970/71–71/72* |
| LIVESTOCK PRODUCTS | | | | | |
| Beef and veal | 132 | 127 | 137 | 167 | 146 |
| Pigmeat | 167 | 185 | 218 | 224 | 248 |
| Mutton and lamb | 2 | 2 | 1 | 2 | 3 |
| Eggs | 90 | 92 | 93 | 93 | 96 |
| Milk[d] | 3.764 | 3.360 | 3.429 | 3.219 | 2.729 |

[a] 1952 only, indices for other two years not available.
[b] Winter rye only.
[c] Includes winter and spring rape, winter and spring turnip rape and white mustard.
[d] Delivered to dairies.

basis (of total population and not only agricultural population), showed no change (Table 12.4). An increase of 50 per cent in the level of crop production was to a large extent offset by a decrease in the level of livestock production. The expansion in crop production involved large increases in barley and oats, with significant decreases in mixed grains and hay and moderate decreases in sugar beet and potatoes. The reduction in livestock production involved a 19 per cent decrease in milk production which more than offset increases in meat production, mostly of pigmeat. Changes in crop acreages and in livestock numbers are summarized in Table 12.5 and indicators of changes in technology (crop yields per acre, milk production per cow, machinery on farm holdings and fertilizer consumption) are contained in Table 12.6.

As measured by farm receipts and expenses, the Swedish agricultural sector expanded during the period under review. Compared with 1953, cash receipts were higher in 1971 by 70 per cent and cash expenses by almost 110 per cent. The generally rising level of prices which characterized this period accounted, however, for a large part of the increase in cash receipts and, perhaps to a lesser extent, in cash expenses. Whereas cash expenses represented about two-thirds of the value of cash receipts in 1953, this proportion had risen to more than 80 per cent in 1971. As measured by the total net cash income from Swedish farms, the dimensions of the agricultural sector in 1971 approximated that of 1953 in terms of current prices but in real terms (constant prices) were significantly smaller. The total net cash income of farm families from non-farm sources was, however, about five times as large in 1971.

The structure of the Swedish agricultural sector has been greatly altered as the reduction in the number of farms has not been spread evenly throughout the sector (Table 12.7). Between 1951 and 1971 the number of farm holdings with 2.1 to 10.0 hectares of arable land was reduced by more than 60 per cent, and of those with 10.1 to 20.0 hectares of arable land by almost 40 per cent. During the same period, there was practically no change in the number with 20.1 to 30.0 hectares of arable land, and an increase of more than a third in the number with more than 30 hectares of arable land. With respect to the production regions, the number of farm holdings in the northern areas was reduced by 55 per cent as compared with only 40 per cent in the plains of southern and central Sweden. The structure of the agricultural sector in 1971, in terms of farm holdings grouped according to their area of arable land, is summarized in Table 12.8 with respect to resources, farm receipts and expenses and farm family incomes.

## AT THE NATIONAL LEVEL

Sweden's national policies of full employment, rapid economic growth and

TABLE 12.5 Crop acreages and livestock numbers

| | *1950–2 average* | *1955–7 average* | *1960–2 average* | *1965–7 average* | *1970–2 average* |
|---|---|---|---|---|---|
| | | | *('000 hectares)* | | |
| | | | | | *1970–71* |
| Wheat | 323 | 356 | 304 | 247 | 257 |
| Rye | 113 | 109 | 83 | 55 | 82 |
| Barley | 115 | 234 | 345 | 559 | 653 |
| Oats | 460 | 487 | 510 | 486 | 555 |
| Mixed grains | 329 | 301 | 232 | 128 | 78 |
| Total cereals | 1340 | 1487 | 1474 | 1475 | 1625 |
| Oilseeds[a] | 166 | 75 | 63 | 85 | 108 |
| Potatoes | 114 | 110 | 100 | 69 | 62 |
| Sugar beet | 53 | 51 | 49 | 41 | 40 |
| Green fodder | 99 | 76 | 59 | 42 | 27 |
| Hay | 1244 | 1139 | 1068 | 937 | 793 |
| Pastures[b] | 215 | 215 | 251 | 254 | 216 |
| | | | *('000 heads)* | | |
| Cattle | 2514 | 2342 | 2429 | 2181 | 1879 |
| *of which*: cows | 1535 | 1334 | 1192 | 935 | 743 |
| Pigs | — | 1597 | 1909 | 1933 | 2177 |
| Sheep | 222 | 141 | 151 | 242 | 333 |
| Poultry | — | — | 9579 | 8310 | 8096 |

[a] Includes winter and spring rape, winter and spring turnip rape, white mustard, linseed and other oilseeds.
[b] Arable land only.

TABLE 12.6 Indicators of changes in technology in the agriculture sector

| | 1950–2 | 1960–2 | 1970–1 |
|---|---|---|---|
| Crop yields per hectare | | (*kilogrammes*) | |
| Wheat | 2.006 | 3.072 | 3.815 |
| Barley | 2.220 | 2.557 | 3.009 |
| Oats | 1.615 | 2.401 | 3.199 |
| Oilseeds | — | — | 1.870 |
| Potatoes | 13.465 | 15.114 | 22.015 |
| Sugar beet | 33.534 | 40.211 | 40.553 |
| | *1950/51–51/52* | *1960/61–61/62* | *1970/71* |
| Milk production per cow[a] | 3.962 | 4.483 | 5.252 |
| | *1951* | *1961* | *1971* |
| Machinery on farm holdings | | (*number*) | |
| Tractors | 69,292 | 148,245 | 163,989 |
| Combines | 8347 | 26,345 | 42,064 |
| Binders | 101,608 | 90,913 | — |
| | *1950/51* | *1960/61* | *1970/71* |
| Fertilizer consumption | | (*'000 tons*) | |
| Nitrogenous | 68 | 106 | 226 |
| Phosphoric | 106 | 104 | 64 |
| Potassium | 54 | 83 | 109 |

[a] Average per cow whose production is recorded.

TABLE 12.7 Distribution of farms[a] according to area of arable land per holding

| *Hectares of arable land* | *1951* | *1961* (*number of farms*) | *1971* | *1951* | *1961* (*percentage*) | *1971* |
|---|---|---|---|---|---|---|
| 2.1 – 5.0 | 95,945 | 66,635 | 30,639 | 34.0 | 28.6 | 20.4 |
| 5.1 – 10.0 | 89,755 | 75,017 | 39,135 | 31.8 | 32.2 | 26.0 |
| 10.1 – 20.0 | 59,790 | 53,446 | 36,685 | 21.1 | 23.0 | 24.4 |
| 20.1 – 30.0 | 17,719 | 18,226 | 17,799 | 6.3 | 7.9 | 11.8 |
| 30.1 – 50.0 | 11,234 | 11,960 | 14,920 | 4.0 | 5.1 | 9.9 |
| 50.1 – 100.0 | 5419 | 5410 | 8132 | 1.9 | 2.3 | 5.4 |
| more than 100.0 | 2325 | 2186 | 2704 | 0.8 | 0.9 | 1.8 |
| Total | 282,187 | 232,920 | 150,014 | 100.0 | 100.0 | 100.0 |

[a] Holdings with more than 2 hectares of arable land.

more uniform distribution of income in a balanced economy have played a particularly important role in changing the position of agriculture in the national economy. Its relative position has declined, during the period under review, as a source of GNP, of employment and of personal income. Although the inclusion of forestry in most Swedish farm holdings

Table 12.8 Distribution of agricultural resources, value of products sold, cash production expenses, and net cash income according to size of farm holding, 1971

| | Hectares of arable land | | | | | | |
|---|---|---|---|---|---|---|---|
| | *2.1–10.0* | *10.1–20.0* | *20.1–30.0* | *30.1–50.0* | *50.1–100.0* | *Over 100* | *All farms* |
| NUMBER OF FARMS | 46.5 | 24.5 | 11.9 | 9.9 | 5.4 | 1.8 | 100 |
| RESOURCES | | | (*hectares*) | | | | |
| Land, arable | 13.6 | 18.0 | 14.7 | 19.2 | 18.3 | 16.2 | 100 |
| Labour[a] | | | (*hours*) | | | | |
| operator and family | 29.7 | 29.6 | 16.9 | 14.4 | 7.5 | 1.9 | 100 |
| hired | 13.3 | 14.2 | 9.4 | 13.8 | 17.7 | 31.6 | 100 |
| Total | 27.6 | 27.6 | 15.9 | 14.3 | 8.8 | 5.8 | 100 |
| Capital assets[b] | | | (*'000 Kr.*) | | | | |
| real estate | 24.9 | 21.0 | 13.9 | 15.9 | 12.9 | 11.4 | 100 |
| non-real estate | 21.7 | 19.5 | 13.2 | 17.1 | 13.8 | 14.7 | 100 |
| Total | 23.3 | 20.2 | 13.5 | 16.5 | 13.4 | 13.1 | 100 |
| VALUE OF PRODUCTS SOLD | | | | | | | |
| Crop products | 4.1 | 9.4 | 11.3 | 21.1 | 26.4 | 27.7 | 100 |
| Livestock products | 13.3 | 22.9 | 18.0 | 20.6 | 14.8 | 10.4 | 100 |
| Forest products | 30.3 | 22.6 | 11.9 | 12.0 | 8.8 | 14.4 | 100 |
| Total | 12.6 | 19.1 | 15.5 | 19.8 | 17.4 | 15.6 | 100 |
| CASH PRODUCTION EXPENSES | 12.1 | 17.4 | 14.5 | 19.6 | 18.3 | 18.1 | 100 |
| NET CASH INCOME | | | | | | | |
| Farm operations | 22.3 | 26.3 | 17.9 | 17.5 | 11.1 | 4.9 | 100 |
| Non-farm sources | 56.6 | 19.9 | 8.1 | 6.4 | 4.8 | 4.2 | 100 |
| Total | 39.6 | 23.1 | 13.0 | 11.9 | 7.9 | 4.5 | 100 |

[a] 1970.

[b] Based on value of capital assets as assessed for tax purposes; real estate includes that used for both agricultural and other purposes; market value of agricultural real estate sold in 1971 was approximately 70 per cent above the assessed value.

complicates the use of conventional national accounts data to measure this relative position (Table 12.9), the decline is nevertheless evident. The national accounts-based measures show 6.4 per cent of the economically active population to have been engaged in farming in 1971 and to have contributed less than 2 per cent of the total GNP. However, this conventional statistical comparison seems certain to overstate the productivity gap between the agricultural sector and the rest of the economy in Sweden. As has been noted previously, a large part of the labour force reported on this basis as engaged in farming is actually employed part time, not only in forestry on the farm holdings, but also outside the agricultural sector. Thus, production on Swedish farm holdings (including the forest land that they contain) probably uses only 3 to 4 per cent of the current total national labour input and probably contributes about 3 per cent of the total GNP. Similarly, with about half of the Swedish farmers also engaged in off-farm employment and approximately half of the total net cash income of farm families obtained from non-farm sources, the dependence of the Swedish population on agriculture as a source of personal income appears also to have declined to about 3 per cent.

Not only has the relative position of the agricultural sector in the Swedish economy changed, but it has also become increasingly linked to other sectors of the national economy. As the scale of farm production enterprises has increased, the individual farms have tended to become more highly commercialized, or market-oriented. An increasing proportion of the farm production has been sold to other sectors. The agricultural sector supplies almost all the primary products processed and distributed by Sweden's food industries, which employ about 80,000 persons. It also supplies about a quarter of the raw materials used by the wood processing industries, which account for about a quarter of Sweden's total exports. The sector has become highly mechanized and, with widespread application of advanced production technologies, dependent on inputs purchased from other sectors. Such purchases in 1970 are estimated to have exceeded 3000 million Kr. (machinery 710m, fertilizers 470m, fuel and lubricants and electric power 275m, livestock feed concentrates 770m, services 820m, etc.).

As incomes have risen, the share of personal consumption expenditures used for food has declined. As the scope and range of services provided by food processing and distribution industries has expanded, the 'farmers' share' of consumer expenditures for food has also declined. As a consequence, the farmers' share of total consumer expenditures for food is estimated currently at about 25 per cent.

## AT THE INTERNATIONAL LEVEL

With respect to adjustments at the international level, a prominent feature

TABLE 12.9 Relative position of the agricultural sector in the national economy in terms of national accounts and census data

| | *1951* | *1956* | *1961* | *1966* | *1971* |
|---|---|---|---|---|---|
| NATIONAL ACCOUNTS | | | (1.000.000 Kr.) | | |
| GNP, current market prices | 39,603 | 55,178 | 78,153 | 122,041 | 181,954 |
| Farm sector[a] product, current market prices | 2004 | 2378 | 2572 | 3024 | 3504 |
| | | | (% of GNP) | | |
| | 5.1 | 4.3 | 3.3 | 2.5 | 1.9 |
| Forestry sector[b] product, | | | (1.000.000 Kr.) | | |
| current market prices | 2026 | 2520 | 2619 | 2845 | 3466 |
| | | | (% of GNP) | | |
| | 5.1 | 4.6 | 3.4 | 2.3 | 1.9 |
| CENSUS | | | *1960* | *1965* | *1970* |
| Economically active population | | | | | |
| Total | — | — | 3,244,084 | 3,449,897 | 3,412,668 |
| *of which*: in farming | — | — | 354,228 | 318,400 | — |
| % of total | — | — | 10.9 | 9.6 | — |
| in forestry[c] | — | — | 83,264 | 67,102 | — |
| % of total | — | — | 2.6 | 1.9 | — |

[a] Excluding farm portion of forestry sector.
[b] Including farm portion of forestry sector.
[c] Including hunting.

TABLE 12.10 Imports and exports of food and agricultural products

| | 1966 | 1967 | 1968 | 1969 | 1970 | 1971 | 1972[a] |
|---|---|---|---|---|---|---|---|
| | | | (million U.S. dollars)[b] | | | | |
| IMPORTS | | | | | | | |
| Food and live animals | 478.5 | 469.5 | 513.0 | 526.1 | 640.4 | 617.0 | 688.1 |
| *of which*: meat and meat preparations | 28.6 | 33.4 | 40.0 | 47.4 | 55.0 | 49.1 | 58.4 |
| dairy products and eggs | 8.6 | 9.3 | 8.9 | 10.1 | 15.5 | 21.1 | 13.1 |
| fish and fish preparations | 51.0 | 50.6 | 73.8 | 61.3 | 74.9 | 77.4 | 92.5 |
| cereals and cereal preparations | 32.5 | 29.9 | 30.2 | 29.1 | 33.2 | 32.1 | 35.5 |
| fruit and vegetables | 153.5 | 149.8 | 159.9 | 175.5 | 195.1 | 189.7 | 210.5 |
| sugar, etc. | 15.6 | 15.7 | 16.7 | 15.1 | 21.2 | 28.4 | 31.6 |
| coffee, tea, etc. | 114.9 | 111.6 | 115.7 | 118.4 | 153.1 | 137.2 | 152.8 |
| feedstuffs | 59.1 | 53.7 | 50.7 | 49.5 | 58.7 | 58.3 | 63.7 |
| Beverages and tobacco | 67.5 | 68.3 | 72.1 | 68.9 | 80.7 | 75.4 | 98.5 |
| *of which*: tobacco | 33.4 | 29.4 | 29.7 | 27.5 | 35.0 | 31.3 | 38.9 |
| Crude materials | 127.0 | 119.9 | 110.5 | 116.2 | 114.1 | 113.4 | 133.6 |
| *of which*: hides and skins | 20.2 | 19.7 | 18.1 | 21.9 | 18.7 | 17.9 | 30.9 |
| oilseeds | 16.6 | 14.0 | 12.8 | 9.4 | 11.3 | 13.6 | 11.1 |
| natural rubber | 13.8 | 11.1 | 9.2 | 12.7 | 11.7 | 9.2 | 9.0 |
| textile fibres | 29.9 | 26.5 | 21.9 | 20.7 | 18.5 | 15.2 | 14.7 |
| Animal and vegetable oils | 22.3 | 21.3 | 22.4 | 23.9 | 38.6 | 43.7 | 38.7 |
| Total imports | 695.3 | 679.0 | 717.9 | 735.1 | 860.5 | 849.5 | 958.9 |

*Table 12.10 (continued)*

| | 1966 | 1967 | 1968 | 1969 | 1970 | 1971 | 1972[a] |
|---|---|---|---|---|---|---|---|
| | | | (million U.S. dollars)[b] | | | | |
| EXPORTS | | | | | | | |
| Food and live animals | 121.2 | 122.2 | 116.1 | 140.2 | 133.4 | 179.5 | 229.9 |
| *of which:* meat and meat preparations | 45.6 | 42.8 | 36.8 | 50.5 | 43.7 | 46.4 | 68.1 |
| dairy products and eggs | 13.2 | 9.1 | 11.3 | 9.9 | 6.2 | 9.1 | 23.4 |
| fish and fish preparations[c] | 8.3 | 7.2 | 8.1 | 8.4 | 8.5 | 14.0 | 15.6 |
| cereals and cereal preparations | 34.5 | 43.2 | 39.0 | 46.5 | 44.9 | 77.1 | 77.5 |
| fruits and vegetables | 8.9 | 9.3 | 8.9 | 8.2 | 11.0 | 10.3 | 13.5 |
| sugar, etc. | 1.9 | 2.8 | 3.4 | 5.5 | 5.6 | 6.2 | 7.5 |
| coffee, tea, etc. | 3.0 | 3.2 | 3.5 | 4.1 | 6.0 | 7.1 | 10.2 |
| feedstuffs | 1.3 | 0.6 | 0.9 | 0.8 | 0.8 | 2.0 | 3.7 |
| Beverages and tobacco | 1.8 | 2.2 | 1.8 | 2.5 | 5.2 | 9.0 | 11.5 |
| *of which:* tobacco | 1.2 | 1.8 | 1.3 | 1.8 | 3.5 | 6.9 | 10.2 |
| Crude materials | 50.9 | 41.8 | 44.3 | 51.0 | 44.9 | 44.7 | 59.7 |
| *of which:* hides and skins | 39.4 | 31.0 | 31.4 | 32.8 | 27.3 | 27.4 | 34.9 |
| oilseeds | 2.0 | 2.3 | 5.2 | 9.8 | 6.8 | 7.9 | 12.7 |
| natural rubber | 2.7 | 2.0 | 1.6 | 1.1 | 1.2 | 0.7 | 0.7 |
| textile fibres | 2.6 | 1.8 | 1.5 | 1.9 | 2.4 | 2.1 | 3.2 |
| Animal and vegetable oils | 12.4 | 14.4 | 13.2 | 12.3 | 15.9 | 21.7 | 21.2 |
| Total exports | 186.2 | 180.5 | 175.5 | 205.9 | 199.4 | 254.9 | 322.3 |
| NET IMPORTS | 509.1 | 498.5 | 542.4 | 529.2 | 661.1 | 594.6 | 636.6 |

[a] Preliminary.

[b] At following exchange rates: 1966–70 \$1 = 5.1732 Kr.; 1971 \$1 = 5.110 Kr.; and 1972 \$1 = 4.763 Kr.

[c] Excluding landings of fish abroad.

TABLE 12.11 Imports and exports of 'egentliga'[a] agricultural products

| | 1966 | 1967 | 1968 | 1969 (million Kr.) | 1970 | 1971 | 1972[b] |
|---|---|---|---|---|---|---|---|
| IMPORTS | | | | | | | |
| Animal products[c] | 230.4 | 248.8 | 284.4 | 333.9 | 397.6 | 385.7 | 369.4 |
| Meat | 116.4 | 133.8 | 165.2 | 196.5 | 224.6 | 204.9 | 234.9 |
| Canned or preserved meat | 32.6 | 38.9 | 41.7 | 48.5 | 59.9 | 46.2 | 43.1 |
| Dairy products and eggs | 44.3 | 48.1 | 46.1 | 52.4 | 80.0 | 107.6 | 62.4 |
| Live animals and other | 37.1 | 28.0 | 31.4 | 36.5 | 33.1 | 27.0 | 29.0 |
| Vegetable products[c] | 604.9 | 527.8 | 511.5 | 481.0 | 608.4 | 573.0 | 574.1 |
| Breadgrains | 58.2 | 31.4 | 29.3 | 19.3 | 17.2 | 9.0 | 8.7 |
| Feedgrains | 22.8 | 22.5 | 23.9 | 25.8 | 27.9 | 17.5 | 15.0 |
| Other cereal products | 75.2 | 85.0 | 86.5 | 93.9 | 100.7 | 104.2 | 108.7 |
| Sugar, syrup, molasses, etc. | 53.3 | 50.3 | 53.5 | 49.5 | 68.2 | 105.8 | 103.9 |
| Dry edible peas, beans and similar | 20.0 | 20.9 | 28.0 | 31.1 | 47.5 | 39.6 | 42.0 |
| Potatoes and products and preparations | 14.4 | 6.8 | 13.4 | 23.2 | 62.8 | 9.3 | 12.9 |
| Copra | 75.4 | 63.5 | 56.2 | 37.8 | 43.8 | 53.8 | 35.0 |
| Oilseeds | 10.3 | 8.7 | 9.1 | 10.1 | 13.4 | 15.0 | 16.3 |
| Oilcake | 243.5 | 213.1 | 186.2 | 167.6 | 198.3 | 190.5 | 206.1 |
| Other vegetable products | 31.8 | 25.6 | 25.4 | 22.7 | 28.6 | 28.3 | 25.5 |
| Fats and oils[d] | 98.4 | 98.2 | 114.3 | 124.9 | 183.5 | 211.8 | 188.4 |
| Total imports | 933.7 | 874.8 | 910.2 | 939.8 | 1.189.5 | 1.170.5 | 1.131.9 |

*Table 12.11 (continued)*

| | 1966 | 1967 | 1968 | 1969 (million Kr.) | 1970 | 1971 | 1972[b] |
|---|---|---|---|---|---|---|---|
| EXPORTS | | | | | | | |
| Animal products[c] | 324.2 | 286.5 | 267.5 | 336.8 | 284.1 | 307.1 | 464.4 |
| Meat | 230.8 | 217.3 | 190.3 | 260.5 | 224.5 | 235.1 | 322.1 |
| Canned or preserved meat | 6.3 | 5.3 | 2.7 | 2.1 | 2.7 | 3.3 | 4.6 |
| Dairy products and eggs | 68.4 | 46.8 | 58.2 | 51.3 | 32.1 | 46.7 | 111.6 |
| Live animals and other | 18.7 | 17.1 | 16.3 | 22.9 | 24.8 | 22.0 | 26.1 |
| Vegetable products[c] | 221.2 | 270.5 | 266.3 | 332.3 | 312.9 | 477.1 | 471.0 |
| Breadgrains | 49.5 | 74.6 | 86.7 | 83.9 | 92.2 | 73.9 | 91.1 |
| Feedgrains | 64.9 | 90.6 | 66.2 | 95.1 | 64.0 | 229.2 | 170.0 |
| Other cereal products | 59.4 | 55.7 | 47.1 | 60.4 | 76.3 | 90.8 | 106.1 |
| Sugar, syrup, molasses etc. | 4.2 | 7.4 | 8.7 | 18.9 | 15.4 | 15.0 | 14.8 |
| Dry edible peas, beans and similar | 5.3 | 3.1 | 2.1 | 2.1 | 0.6 | 1.2 | 2.0 |
| Potatoes and products and preparations | 1.0 | 1.9 | 1.1 | 1.0 | 1.6 | 4.6 | 5.1 |
| Copra | — | — | — | — | — | — | — |
| Oilseeds | 10.4 | 11.8 | 27.0 | 50.9 | 35.4 | 40.4 | 60.7 |
| Oilcake | 1.9 | 1.5 | 1.7 | 0.9 | 1.0 | 1.6 | 4.9 |
| Other vegetable products | 24.6 | 23.9 | 25.7 | 19.1 | 26.4 | 20.4 | 16.3 |
| Fats and oils[d] | 53.9 | 64.4 | 61.8 | 55.3 | 68.9 | 94.4 | 90.2 |
| Total exports | 599.3 | 621.4 | 595.6 | 724.4 | 665.9 | 878.6 | 1025.6 |
| NET IMPORTS | 334.4 | 253.4 | 314.6 | 215.4 | 523.6 | 291.9 | 106.3 |

[a] Essentially products subject to price regulation measures in Sweden which are those actually produced domestically. [b] Preliminary.
[c] Excluding fats and oils. [d] From animal and vegetable sources.

of the Swedish experience has been the largely passive role envisaged for changes in the volume and composition of its foreign trade in food and agricultural products. Although these adjustments have reflected changes in Sweden's self-sufficiency in food and feed, they have been mostly the consequences of changes in the available supplies of domestically produced farm products. In short, foreign trade has adjusted to domestic production rather than vice versa. During the period under review, Sweden's balance of payments position has generally been sufficiently strong not to have encouraged special measures to expand exports or to restrict imports of food and agricultural commodities. The level of per caput food consumption has been relatively high throughout the period and the increase in food requirements has, consequently, been limited largely to the relatively small increase in population and to changes in the composition of the food consumed.

Although Sweden's price and related foreign trade measures stabilized domestic prices for farm products at higher than world market levels, these higher prices have been designed primarily to support Swedish farm incomes and to assure maintenance of the desired minimum domestic food production capacity rather than to restrict imports. In view of the relatively high level of food consumption in Sweden, these higher prices appear not to have reduced the total demand for food in Sweden, at least not to any important extent. Priority in the fulfilment of these requirements has, however, been given to domestic production and the higher prices have tended to provide incentives for this domestic production to be increased. With respect to those agricultural products for which Sweden's exports have increased, this has been not the result of policy goals or active promotional efforts, but rather the unsought consequence of Swedish production exceeding domestic requirements. The financial losses of these exports have been borne largely by the Swedish farmers who have produced the commodities that have had to be exported.

Throughout the period under review, Sweden's foreign trade in food and agricultural products has shown a large net import balance (Table 12.10). As might be expected from its northern location, a large part of these Swedish imports have consisted of commodities not produced domestically; coffee, tea, cocoa, fruits and vegetables, tropical vegetable oils, cotton, natural rubber, etc. With respect to the major agricultural commodities produced domestically and subject to government price regulation, Sweden's foreign trade has, however, also shown a net import balance (Table 12.11). This balance declined sharply in 1971 and again in 1972, as a result of increased exports of feedgrains, meat and dairy products.

NOTES

1 Includes income of the operator and members of his family from employment on farms other than his own and in forestry not included in his farm holding.
2 Weights used in the construction of these indices are regional wheat-based price relatives during the base period 1961–5.

# 13 Issues, Options, and Implications[1]

Basic policy issues involved in Sweden's agricultural adjustment experience have included the reasons for support to the agricultural sector, the extent to which this support should be provided, and the forms that it should take. These issues, and the options that have been available for Sweden, correspond generally to those for other high-income, industrialized countries. Agricultural adjustment appears, however, to have proceeded further in Sweden than in other developed countries. Therefore, the Swedish experience seems likely to contain implications concerning these issues and options that may not, as yet, have become so clearly evident in other of these countries.

Reasons for the Swedish Government's policy of support for its agricultural sector have been reflected in its major agricultural policy goals: in short, production, incomes and economic efficiency. During the period under review, the priorities accorded to the related policy goals have shifted. These shifts have reflected mainly changes in available technology, evolution of Sweden's national economic and social goals, and developments in the Swedish economic situation and social conditions, particularly as these have combined to bring increasingly close integration of the agricultural sector into the national economy and, during more recent years, to give rise to concern about environmental conditions.

The long-term SSR production goal has been – and continues to be – the most important reason for the Swedish Government's support of its agricultural sector. The need to maintain a domestic production capacity adequate to assure fulfilment of the country's food requirements in time of emergency has continued to be accepted in Sweden as a requisite for maintenance of the country's traditional position of neutrality.

During the period under review, distinctly less importance has come to be attached to the incomes of farm families as a reason for support to the agricultural sector. As the sector came increasingly to be treated as an integral part of the national economy, the incomes of farm families came to be considered as primarily the concern of national income policies, rather than as a reason for support to the agricultural sector. The 1967 guidelines, based primarily on economic considerations, adopted this position explicitly, and the means for improving incomes for farm families became generally the same as those applied for families in other sectors of the economy.

The importance attached to economic efficiency in the agricultural sector, as a reason for support to it, has shifted significantly during the period under review. Initially, higher levels of efficiency in farm production tended to be considered primarily as a means for increasing the incomes of farm families. As the economy expanded during the 1950s and demand for labour in industrial and service sectors rose, it became increasingly evident that – from the macroeconomic point of view – rapid economic growth depended not only on full employment but also on ever-higher levels of productivity in all sectors of the economy. Thus the purpose of improved efficiency on farms became increasingly to release labour for transfer to more productive employment in other sectors. This need for further transfer of labour to other sectors appears, however, to have become less urgent as the rate of Sweden's economic growth has slackened ever since 1970. In any case, the continued rapid movement out of agriculture since the end of the Second World War has sharply reduced the number of workers available for transfer to other sectors as the result of further improvements in efficiency on farms.

These major policy goals have been intricately interrelated. The implication of the production goal that the level of total agricultural production was too high distinctly limited the number of farm families for which the income goal could be attained from farming operations, if producer prices (or returns) were not to be raised to unacceptably high levels. The production goal also implied that the aggregate of resources engaged in agricultural production was too large, which distinctly limited the number of farms that could be accommodated within the sector, if they were not to be of such small sizes as to preclude application of improved production and management technology. The implication of the economic efficiency goal that the level of this technology should be raised also limited the number of farms that the sector could accommodate, if the aggregate of agricultural resources were not to be increased and total production, consequently, expanded. Thus, contraction of its agricultural sector (in terms of both resources and production) became Sweden's increasingly explicit strategy for simultaneous attainment of its major agricultural policy goals.

## ISSUES AND OPTIONS

With respect to the production goal, basic policy issues[2] continue to be: how large should Sweden's total agricultural production be? What should be the commodity composition? The answer to the first of these has been provided by the SSR-based production goal, generally accepted in connection with 1947 guidelines and explicitly adopted in those of 1967:

the total production should be that which would result from efficient utilization in peacetime of the aggregate of resources needed to maintain a domestic production capacity adequate to assure fulfilment of the country's food requirements in time of emergency. The guidelines for Sweden's agricultural policies have, however, not provided a firm answer to the second of these questions.

A secondary policy issue relating to the volume of the production capacity needed for security reasons concerns the size of the stocks of food commodities and agricultural production requisites; the larger these stocks, the smaller the production capacity needed. A further policy issue relates to the aggregate of resources needed to provide a given production capacity and concerns the level of technology in the agricultural sector; the higher the level of productivity, the smaller the aggregate of resources needed. A related policy issue involves the composition of the aggregate of resources to be engaged in agriculture, the area of land or amount of capital per person employed in the sector, etc.

The first of these secondary policy issues (i. e. the size of reserve stocks) lies outside the scope of the present study. The Swedish option with respect to the second has continued to be expressed by the economic efficiency goal (see below) which has precluded restraints on improvement of production and management technology. Concerning the third, the Swedish option has been clearly expressed only with respect to the arable land – man ratio, as operational aims of the CAB farm rationalization activities have evolved from the creation of 'basic farms' (10 to 20 hectares of arable land) to 'norm farms' (20 to 30 hectares of arable land) and, under the 1967 guidelines, to efficient farms regardless of size. The amounts of capital involved in the special rationalization farms[3] might be interpreted as representative of the Swedish option with respect to the capital – man ratio, at least in northern Sweden.

In the 1967 guidelines, Sweden clearly opted for maximum reduction in the size of its agricultural sector, subject only to the overriding security-based consideration. The SSR calculation in connection with these guidelines demonstrated that the Swedish agricultural sector was larger than needed for this purpose and could therefore safely be reduced. It was also accepted that a reduction in the agricultural sector was highly desirable (and indeed essential) in order to release labour for transfer to other, more dynamic sectors of the economy. A smaller agricultural sector was also considered desirable to minimize the total support required and thereby to lessen the burden on the economy, in the form of either higher consumer prices for food or of larger expenditures from the government budget.

To bring about the desired contraction of its agricultural sector, Sweden has opted generally for indirect rather than direct methods. Area, production or marketing allotments or quotas have been considered as likely to encounter serious administrative difficulties and to require direct and centralized government interventions, such as would not be expected to prove acceptable under Swedish conditions. It has moreover been considered that such allotments or quotas would, in any case, tend to delay the desired contraction of the agricultural sector by encouraging labour and other resources, which otherwise would transfer to alternative uses, to continue in agriculture. Thus Sweden has sought to attain the desired contraction of its agricultural sector mainly by the expansion of opportunities for attractive alternative employment outside the agricultural sector. Farm rationalization activities of the CABs have been designed especially to facilitate the transfer out of agriculture of established farmers and to ensure that the outflow of labour was accompanied by suitable adjustments in the use of agricultural land.

Although the aggregate of resources engaged in agriculture has been reduced rapidly and a zero rate of growth in agricultural production per caput achieved, Sweden's total agricultural production remains in excess of the long-term SSR production goal. An important reason for this is that the contribution of improved technology to the expansion of agricultural production has been greater than was foreseen and, consequently, the volume of resources no longer needed in agriculture has proved to be larger than was generally anticipated. More vigorous and sustained use of changes in producer prices as a means for restraining production would have been necessary to hasten the contraction of the sector's resources.

The rapid transfer of resources out of the sector that has taken place has, however, served to reveal a number of constraints on such adjustments. The most dramatic feature of this transfer has of course been the sharp reduction in the agricultural labour force. The demand for additional labour in industrial and service sectors has slackened since 1970, perhaps temporarily. Also, the need to maintain a certain population in rural areas for environmental reasons has become more explicit during recent years. Moreover, the movement of labour out of agriculture that has already taken place has been a highly selective process; those who have chosen to move to other employment have tended to be those who had the most to offer in the way of contributions in other sectors and whose transfers were, therefore, most advantageous from both personal and macroeconomic points of view. An increasing proportion of the decreased number remaining on farms has come to consist of older, less adaptable, less mobile persons whose transfers to other employment would likely prove not only to be personally unpleasant but also costly to the economy for the additional housing and infrastructure services that would be required to accomodate them in other locations. The limited type of rationalization support (B-support)[4] that has been available since 1971 was designed with such cases, *inter alia*, in mind.

Compared with the sharp reduction in the agricultural labour force, the reduction in the area of arable land has been more modest. However, with continued improvements in technology and increases in crop yields per hectare, the area of arable land no longer needed for crop production seems certain to increase, if the SSR production goal is not to continue to be exceeded. If supplies for export are to be avoided, an important reduction in the area presently devoted to cereals appears to be required. Possibilities for shifting such land to other crops are quite limited, however, except possibly to sugar beet or oilseeds. With present price and cost structures, the profitability from technically suitable alternative uses (such as forestry) frequently proves to be significantly less than that from current uses. In any case there are reported objections, in at least some areas, to widespread replacement of open fields with forests on the basis that this will 'spoil the landscape'.

With respect to the income goal, the obvious policy issue continues to be: what should the average level of incomes for farm families be? The general answer, supplied initially by the 1947 guidelines, has been that farm families should have the same opportunities as other groups of the population to obtain reasonable incomes. Under these guidelines, Sweden opted for stabilization of domestic market prices for major farm products at higher than world market levels as the means for providing farm families with such income opportunities.

As the Swedish economy expanded at a relatively rapid rate during the 1950s, large numbers of persons engaged in agriculture found attractive opportunities for alternative employment in industrial and service sectors. At the same time application of improved production and management technology on farms reduced labour requirements there. As these trends continued during the 1960s, it became increasingly accepted that transfer of labour out of agriculture offered a more advantageous means for improving the income position of farm families than did support to the agricultural sector. The 1967 guidelines reflected this view, confirming that while farm families should have the same income opportunities as other groups no special policy goal was needed. For improvement of the income position of families remaining on farms, these guidelines looked primarily outside the agricultural sector, to the transfer of families with small, non-viable holdings to other full-time employment, to the expansion of off-farm employment on a part-time basis, etc. An increasing proportion of the families that have remained on farms have opted for the latter, i. e. to become essentially part-time farmers. Income from off-farm sources has become a relatively large part (slightly more than half in 1971) of the total income of farm families.

The form in which support should be provided to the agricultural sector has, however, been at least as prominent a policy issue as the level of the support. Throughout the period under review, Sweden has chosen to provide most of the support in the form of higher producer prices. The 1967 guidelines confirmed in principle the so-called high price line, under which producer prices are supported at higher than world market levels and the cost of such support passed on to consumers as part of the retail prices.

These guidelines directed, however, that the feasibility of changing the so-called low price line should be reconsidered after a period of years; under this arrangement, producer prices would be linked to world market levels and support to domestic producers would be in the form of deficiency payments from the government. A major item in the terms of reference for the 3rd Royal Commission is reconsideration of this feasibility. Full application of the high price line was suspended, at least temporarily, when the government froze consumer food prices from 1 January 1973. The increase in producer prices which was due at that time (under the provisions of the 1971–4 price regulations) was covered by deficiency payments from the government budget.

The choice between the high price line and the low price line does not *per se* affect the extent to which the support may provide incentives for increasing production or for retaining resources in agriculture. These incentives are related to the amount of the support rather than to the form in which it is received by the domestic producers, and the choice between the two systems concerns only the latter. This choice does however have serious implications for the level of retail food prices and the case for changing to the low price line stands mainly on this basis. Although the total cost of the support for the economy would not necessarily be altered by a change to the low price line, the burden of this cost would almost certainly tend to be shifted from lower-income to high-income segments of the population. On this basis, the major issue involved in this choice probably relates more directly to national income distribution policy than agricultural policy.

Although a major part of the support provided for the Swedish agricultural sector has continued to be in the form of higher producer prices, the reasons for providing support in this form have shifted to include considerations other than the income position of farm families. The desire to avoid an excessively rapid 'flight from the land' has become an important consideration. During the 1960s the rapid movement of population out of the northern and other more remote areas posed serious problems for maintenance of public services and threatened disruption of the local social and commercial infrastructures. During recent years the need to maintain an active farm population in these areas (on the basis of security and environmental considerations) has also become an important reason for support, including the special price supplements for milk and since 1971 for meat, for farms in northern Sweden.

With respect to the *economic efficiency* goal, the equally obvious policy issue would seem to be: what should be the average level of productivity of resources devoted to agricultural production? The agricultural policy guidelines and the national policy goals of full employment and a rapid rate of economic growth imply that resources engaged in agriculture should be as productive as those employed in other sectors of the economy. This response, reflecting essentially economic considerations, has however not been the complete answer to this question for the Swedish agricultural sector. Throughout the period under review, considerations of other than an economic nature have continued to be involved.

The guidelines of 1947 were based mainly on social considerations and accorded higher priority to improving farm incomes than to increasing agricultural productivity. The SSR production goal has of course been based on security considerations throughout the period. The Swedish policy which, until 1965, sought to restrict ownership of agricultural land to active farmers also clearly reflected non-economic considerations. Although economic considerations seem to have been given greater weight during the 1960s, maintenance of the environment has become an additional important consideration during recent years. As with the income goal, Sweden seems to be opting for a two-tier agricultural sector with respect to economic efficiency; one composed of a relatively small number of large-scale farms on which the productivity of resources should be comparable to that in other sectors, and the other composed of part-time farms, 'guardians of the environment', etc., on which the productivity of resources in terms of agricultural output would not be expected to be comparable to that on the large-scale farms and in other sectors.

As the principal means for increasing efficiency in the agricultural sector, Sweden has opted for the rationalization of farm structures and operations. The government established the CABs with responsibility for administration of its support for this rationalization and delegated operational authority to the individual Boards. This arrangement has enabled flexibility in the CAB operations and served to avoid centralized administration of the rationalization support, such as might have been distasteful to the people in rural areas who were affected. Under this arrangement, the CAB activities have evolved to include a comprehensive array of financial assistance to individual families who have chosen to remain on farms (or to become farmers) to enable them to develop economically viable production units. The Boards have also provided planning and advisory services for families who desired to leave agriculture as well as for those who have chosen to remain.

The support provided through the CABs has assisted the creation of larger-scale and more efficient farms. These prove generally to be more profitable for the operators, and the direct benefits from support provided in this form accrue, therefore, mainly to those families whose farms undergo rationalization. However, consumers also benefit from this support. As was emphasized in the terms of reference for the 3rd Royal Commission, continued rapid rationalization of Swedish agriculture offers the best guarantee that prices for farm products become as low as possible (subject to the overriding SSR production goal).

## IMPLICATIONS OF SELECTED FEATURES OF ADJUSTMENT

In most respects, the results of Sweden's agricultural adjustment experience have been not unlike those in other high-income industrialized countries. Since agricultural adjustment appears, however, to have reached a more advanced stage in Sweden, certain features of the Swedish experience seem likely to be particularly relevant for other of these countries as agricultural adjustment proceeds there.

### *1 The role of national economic and social policies and programmes*

The goals of Sweden's national economic and social policies have included full employment, a rapid rate of economic growth and a more uniform distribution of incomes in a balanced economy. These policy goals and the programmes for achieving them have constituted a crucial part of the setting for Sweden's agricultural adjustment experience. Without them Sweden's agricultural adjustment could hardly have proceeded, at least not in the manner and to the extent that it has. Within the framework provided by them, it has been increasingly accepted that the agricultural sector should become fully integrated into the national economy. As a consequence, the programmes and measures have evolved with this purpose clearly in mind. Thus they have been designed not simply to treat the agricultural sector 'the same as any other sector in the economy', but to deal with its special characteristics which have in the past tended to hamper or to delay this integration. Broadly speaking, it might be said that Sweden's agricultural policies have increasingly treated these special characteristics of the agricultural sector with special measures, and differentiated them from the general policies for agriculture which have been formulated on the basis of the needs of the economy as a whole.

### *2 The self-sufficiency ratio as an instrument of agricultural policy*

Assurance of supplies to meet national requirements for food and other agricultural products, from either domestic production or from imports, is of course a fundamental responsibility of a national government. This responsibility has, however, received more explicit attention in Sweden, as a reason for government support of the agricultural sector, than in most other high-income, industrialized countries. Although this responsibility is rarely ignored by any national government, at least for long, it has tended not to receive such widespread public attention and the policy options available not to be so fully examined as has been the case in Sweden.

The SSR, as calculated for Swedish policy purposes, has served to assure the government and others concerned with public policy that the food production capacity of Sweden's agricultural sector was not smaller than needed to ensure fulfilment of the country's food requirements in time of emergency. Since the sector's capacity has remained larger than needed for this purpose, the long-term SSR production goal has, from a day-today operational point of view, played a limited role in Sweden's agricultural programmes. It has, nevertheless, served to establish the direction of the adjustments desired in the size of the sector, in terms of both production capacity and resource requirements, and to assure that contraction of the agricultural sector was not inconsistent with Sweden's national interests. While the SSR has had no direct operational relationship with the level of Sweden's imports of food and agricultural products since the end of the Second World War, its existence has made acceptable the contraction of the agricultural sector. In turn this has meant that domestic production of these commodities has certainly been lower than it would otherwise have been and, consequently, that the level of imports has been higher and the volume of exports lower.

The Swedish experience has, nevertheless, served to confirm that a technically sophisticated application of the concept of self-sufficiency in food is required for it to serve as an effective instrument of agricultural policy. The usefulness of an SSR calculated on the basis of domestic production and of national consumption both aggregated in terms of calories seems certain to be limited mainly to emergency situations in which calories are the primary consideration. In other situations, the commodity composition of the food as well as its nutritional content (in terms of protein from both animal and vegetable sources, fat, vitamins, minerals, etc.) would need to be taken into account. Swedish experience also indicates that for countries that import agricultural production requisities (petroleum products, fertilizers, pesticides, other agricultural chemicals, and feedstuffs, machinery and equipment, etc.), concepts of 'gross' and 'net' SSRs would also be needed.

### *3 Economic considerations involved in contraction of the agricultural sector*

With its agricultural sector larger than needed on the basis of considerations of security and efficiency, the Swedish policy option to seek contraction of the sector has been based principally on considerations of an economy nature. The decision has reflected, in part, Sweden's northerly location and other relatively unfavourable natural conditions which tend to limit the country's potential for agricultural production. However, it has also reflected, to a major extent, explicit acceptance of the view that Sweden's comparative advantage is generally greater in sectors other than agriculture except possibly at times of temporarily high prices for agricultural products on world markets.

More specific considerations involved in the Swedish options have included, principally, the need for additional labour in industrial and service sectors to enable a rapid rate of economic growth in the national economy. Also involved has been the need to make additional land available for enlargement of the farms that remained in order that they could be transformed into units large enough to be economically and technologically viable. From a practical point of view, an important consideration has been the need to avoid production of specific farm products in excess of domestic requirements, since financial losses would be incurred on the export of surpluses.

*4 Foreign trade as an instrument for agricultural adjustment*

Because of the largely residual role that Sweden's policies have provided for foreign trade in food and agricultural commodities, the experience with agricultural adjustment at the international level has been quite limited.

If the Swedish Government were to opt for an increase in imports of competing food and agricultural products from developing countries, its agricultural adjustment experience since the end of the Second World War suggests that major constraints would involve the security-based long-term SSR goal, the environmental-based aim to maintain a certain population in rural areas, and suitable alternative uses for the arable land no longer needed for crop production.

Given Sweden's high levels of per caput consumption and its low rate of population growth, the rate of increase in Sweden's total requirements for food and agricultural products is expected to be relatively low. Shifts in the composition of the requirements are, however, expected as consumers' incomes continue to rise and their tastes to change. It is to be anticipated that these shifts will have greater effects on the form in which consumers purchase their food (more highly processed, more completely prepared, convenience foods, away-from-home meals, etc.) than on the commodity consumption of the raw materials required at the farm production level.

Sweden applies no foreign trade measures designed specifically to restrict imports of non-competing food and agricultural products. Consequently, there would appear to be slight opportunity for Swedish imports of these commodities (coffee, tea, cocoa, tropical fruits, cotton, natural rubber, etc.) to be expanded. There might, however, be opportunity for increased imports of certain tropical fruits if, by improved handling methods and organization of marketing, prices could be reduced to remove them from the category of luxury items.

Thus, increased Swedish imports of food and agricultural commodities from developing countries would be likely to consist largely of competing products and to displace either domestic production or imports from developed countries. If displacement were limited to domestically produced supplies, with imports from developed countries maintained at present levels, it appears that imports from developing countries might be increased for wheat, rye, barley, oats, potatoes, a limited variety of fruits and vegetables, sugar, a limited variety of vegetable oils and oilseeds, meat, dairy products and eggs. If, however, the possibility for displacing imports from developed countries were also included, the list could be expanded to include maize, soybeans, a greater range of other oilseeds and vegetable oils, citrus fruits, a greater variety of fruits and vegetables, etc.

In any case, the security-based long-term SSR goal would limit the extent to which domestic production might be reduced to accommodate increased imports. To the extent consistent with this goal, the most evident options available to the Swedish Government would be direct restrictions on domestic production of the commodities of which increased imports from developing countries were desired, or downward adjustments in the relative levels at which domestic producer prices for these commodities were stabilized to provide incentives for Swedish farmers to reduce their production. The first would represent the more drastic departure from the basic principles that have governed the evolution of Sweden's agricultural adjustment policies since the end of the Second World War. The second would have important implications for appraisal of the economic viability of farms in connection with the CAB rationalization activities, since the level of producer prices is a critical element in these appraisals. Thus it would be likely to affect the arrangements needed to maintain, for environmental reasons, a certain population in rural areas.

A system of special preferences for imports from developing countries would need to be introduced to avoid the additional imports being supplied by developed countries. This would mean abandonment of the traditional Swedish policy of non-discrimination between sources of imports of food and agricultural commodities. For example, the desire to increase Swedish imports from developing countries was a consideration in the 1967 decision to restrain domestic sugar beet production. With non-discrimination between sources, an important part of the subsequent increase in sugar imports have in fact come from developed countries, although developing countries have probably been amongst the original sources of the raw sugar required to provide sugar imported from developed countries.

NOTES

1 It is perhaps appropriate to recall that the selection of the material included and the interpretations and comments contained in this report are entirely the responsibility of FAO. This applies particularly to the issues, options and implications contained in the section that follows.
2 In addition to these policy issues, there have been and continue to be, a wide range of technical issues relating to definitions, estimating procedures and methodology used in calculating the SSR goal.
3 See 'Farm Rationalization Measures', p. 163.
4 See 'Special Measures', p. 167.

# 14 Summary of Major Conclusions

Sweden's agricultural adjustment experience is an undoubted success story from the point of view of Swedish national goals. The agricultural sector has become a more fully integrated part of the national economy and approximate income parity has been attained as between families in the farm and the non-farm segments of the population. A zero rate of growth in Swedish agricultural production per caput has been achieved. Between 1951 and 1971 the agricultural labour force was reduced by almost 75 per cent. Productivity in the agricultural sector was increased and the profitability of individual farms rose with application of improved production and management technology. Although much has been accomplished, agricultural production capacity remains above the target level, a large number of relatively small farms remain, and further improvements in technology and increases in productivity are to be anticipated. Thus, agricultural adjustment in Sweden is not a completed process, as indeed it cannot realistically be expected ever to be in a country with a dynamic economy.

The Swedish experience illustrates how a high-income, industrialized country, with developed institutions and a tradition of administrative efficiency, can bring about structural adjustments in its agricultural sector without excessive hardships to the people directly affected or to the nation as a whole. In this respect, major ingredients of the Swedish experience have included the maintenance of a high rate of growth in the national economy, which has provided attractive opportunities for alternative employment and thus labour has been pulled (rather than pushed) out of farming; generous transfer assistance to people who have desired to leave farming; a wide range of services and assistance to those who desired to remain in farming, or to become farmers, to enlarge and to equip their farms adequately for effective application of available technology; and the provision of assured and relatively stable markets for farm products, which has both enabled and encouraged farmers to plan their investments and otherwise to rationalize their operations.

Except at the international level (as noted below), Sweden's agricultural sector appears no longer to constitute a permanent 'special case'. A small number of relatively large and efficient full-time farms produce the bulk of Sweden's agricultural output and, within the present price and income distribution structure, are mostly commercially viable units. A considerably larger number of farms, although too small to attain comparable levels of efficiency, are also mostly viable because of family income from off-farm sources. Thus it is possible for Sweden's agricultural policies and programmes, as for example the levels negotiated for producer prices, to be related primarily to the first group and not to be distorted to provide income support for the second group.

These programmes and measures have, however, encountered a number of constraints. Improved technology has increased productivity to such an extent that the outflow of labour and other resources has been more than offset; as a consequence, total farm output has not been reduced as intended and for some commodities (such as wheat) production has, in fact, risen to create surplus problems. On account of broad national consideration, it has been desired to maintain a certain agricultural population in remote and unfavourable areas where farm production is admittedly a high-cost proposition. A major constraint on acceleration of the movement of farm families out of agriculture has been the difficulties experienced in maintaining public services and commercial and social infrastructures for the reduced population that remained in rural areas. For this and other reasons it has not been considered feasible to make sustained and vigorous use of producer prices as a means for restraining production further. It has also, for a variety of reasons, not been considered feasible to apply direct production controls for this purpose.

Estimates of the total cost of Sweden's agricultural programmes tend inevitably to be controversial because of the manner in which these costs have been distributed. They have been borne largely by the consumers directly, in the form of higher food prices. The actual impact of these programmes on consumer food prices has, however, been obscured by the fact that during the same period processing and distribution costs have contributed increasingly to the rise in retail food prices. Price support and stabilization measures (including the subsidization of exports) have been financed mainly from levies on imports and on domestic producers and processors. Costs borne by the government budgets have consisted mainly of grants and administrative expenses under the programmes for rationalization of farm structures and operations and for assisting farm families to transfer out of agriculture, and of supplementary commodity price payments to producers in northern Sweden.

The success of Sweden's agricultural adjustment effort has undoubtedly been facilitated by the establishment of a nationally accepted self-sufficiency ratio for food. This indicated the direction and, although less precisely, the extent of the change that was desired in the country's food production capacity. With actual production well above the level stipulated by the SSR, policy measures for rapid transformation of the farm sector could be undertaken with assurance. Despite the fact that it has not been directly invoked for the day-to-day application of Sweden's agricultural policy measures, it seems fair to say that in the absence of the SSR goal, Sweden's agricultural adjustment could not have proceeded so rapidly. Without agreement that Sweden's agricultural sector could be contracted without endangering fulfilment of the country's food consumption requirements in time of emergency, stronger pressures would almost certainly have been encountered for expansion of production as a means for increasing farm incomes. Experience has demonstrated, however, that a simple calorie-based SSR is not a sufficiently sophisticated instrument for agricultural policy purposes generally.

The Swedish contribution to agricultural adjustment at the international level has been mainly in the form of restraints on its exports rather than of increases in its imports. The largely residual role accorded to foreign trade in agricultural products is in sharp contrast to the strong international orientation of most of Sweden's other economic sectors. In this respect, the agricultural sector continues to constitute a 'special case' for Sweden. Because of the overriding SSR objective – the agricultural component of Sweden's comprehensive security policy – Swedish imports of competing agricultural commodities (whether from developed or from developing countries) remain relatively small. If Sweden's agricultural programmes were to bring the level of domestic production down to that specified by the SSR goal, these imports would likely be somewhat larger. The more important effect of such a reduction in domestic production would, however, be to limit the volume of Swedish exports of these commodities.

# Annex: Swedish Agricultural Institutions and Organizations

The structure of institutions and organizations that has been involved in Sweden's agricultural adjustment experience has included both governmental and non-governmental bodies. The policy guidelines and related legislation has been adopted by the parliament, on the basis of proposals presented by the government. Within the government, the formulation of policies and programmes relating to agriculture, forestry and fisheries has been the responsibility of the Ministry of Agriculture. From an operational point of view, the administration of the programmes has, however, been decentralized, with responsibility at the national level assigned to a series of National Boards (governmental bodies) which function in a relatively autonomous manner under the direction and guidance of the Ministry.

The *National Agricultural Marketing Board* (NAMB) has responsibility for administration of the government's price regulating activities, the price and related foreign trade measures. The NAMB is appointed by the government (Minister of Agriculture) and consists of a chairman and five other members, of which two represent consumers' interests and two represent producers. The price regulating activities are, however, largely carried on by special Marketing Associations, non-governmental bodies organized by commodities, and by farmers' cooperative associations, also non-governmental bodies organized usually by commodities.

The special *Marketing Associations* are independent bodies, functioning under general guidance from the government, with their chairmen (except for the Dairy Association) appointed by the government (Ministry of Agriculture). They include associations for the marketing of milk and dairy products, meat, eggs, cereals, oilseeds, potato starch and sugar. For most of the associations, membership is open to all producers, processors and wholesalers, including farmers' cooperative associations. The objective of each of the Marketing Associations is stabilization of the domestic market for its commodities. For this purpose, they may apply stockpiling schemes and, if necessary, subsidize the export surpluses. They are not, however, entitled to prescribe compulsory use of domestic production. For certain commodities (breadgrains, oilseeds and potatoes for industrial purposes), they also establish 'redemption prices', prices which the association guarantees to producers for all quantities that have not been sold as of a specified end-of-season date. These redemption prices serve the same purpose as prices guaranteed by the government, although the associations rather than the government bear the risk. Each Marketing Association has a compensation fund to cover the cost of its stabilization activities. These funds are derived from marketing and/or processing fees collected from the producers or processors of the commodities involved, from compensatory import fees collected on imported supplies of commodities for which domestically produced supplies are subject to marketing and/or processing fees, from the fixed or variable levies collected on imports of the products that it handles, and, in certain instances, from government grants. The scope and magnitude of the association's market stabilization activities has tended to be limited by the availability of their compensation funds.

The *National Agricultural Board* (NAB) has responsibility for administration of the government's activities for the rationalization of farm structures and operations. The NAB is appointed by the government (Ministry of Agriculture) and consists of a chairman plus at least five other members. At the national level, the NAB works in collaboration with the National Board of Forestry and the National Land Survey Board. At the county level, it supervises and coordinates the work of twenty-four county agricultural boards (CABs), one in each county. The NAB is divided into subject matter departments and special working units. The Swedish Agricultural Information is attached to this board.

The *County Agricultural Boards* were initially established in 1948. Each of them is composed of nine members, four (including the chairman who is usually the governor of the county) appointed by the government, four by the county council, and the director of the CAB. Although appointed as political representatives, some of the members of the CABs are usually farmers or persons with connections to agriculture. Major questions of principle and important cases are decided by the CAB in plenary sessions. Each CAB consists, however, of two permanent delegations, one for structures and one for production, which make most of the decisions. The structure delegation has up to seven members and the one for production up to five. In some cases there are in addition special delegations, such as for horticulture, with up to five members. In each rural district (commune) there is a representative of the CAB to provide local contacts with farmers. A programme council and a fisheries council are attached to each CAB, and various joint committees may be formed to maintain contacts with farmers' organizations and other interested groups. The CABs work in coordination with the County Forestry Boards, the County Labour Market Boards, the county land survey offices and the county veterinary officers.

Effective from 1 July 1967, the advisory services, financial assistance and other governmental measures to promote the improvement of Swedish agriculture were brought together under the new CABs, which at that time took over the responsibilities of the former County Agricultural Boards and the government-assisted activities of the County Agricultural Societies. Since that time, the CABs have provided an integrated system of activities focused on farm rationalization, which has included planning and advisory services, application of the provisions of the Land Acquisition Act of 1965, purchase and sale of farm real estate, the administration of the government's support for investment in agriculture (government grants and guaranteed loans from usual credit sources).

A special *Land Fund* was established by the government in 1948 to finance the CABs' land purchase operations. As the scale of these operations expanded and the area of land held by the CABs increased, the Land Fund was enlarged during the 1960s, from 68 million Kr. in 1960 to 277 million Kr. in 1972. Profits are not sought on the land sales but have occurred more often than losses as a result of rising land prices and of growth on forest lands held by the CABs. Interest is not calculated on the capital invested in land held by the CABs nor are administrative costs charged as expenditures from the Land Fund. Real estate taxes and land survey costs related to the farms traded by the CABs are, however, paid from this fund. The difference between expenditure and income for each farm that passes through CAB ownership is covered from the funds appropriated by the government for farm rationalization purposes and does not affect the size of the Land Fund.

The *Federation of Swedish Farmers* was created in 1971 by the merger of the National Farmers' Union and the Federation of Swedish Farmers' Associations. The National Farmers' Union, established in 1929, was a politically oriented organization modelled to a large extent on Swedish trade unions, which sought to defend the interests of farmers as against those of other groups. The Federation of Swedish Farmers' Associations, created in 1940, was the top-echelon organization of the Swedish farmers' cooperative movement. Thus, the new organization represents both the trade union and the economic interests of farmers. Its structure consists of county and district federations and local branches.

*Farmers' cooperative associations* have been a characteristic feature of Swedish agriculture since the nineteenth century. It is estimated that 98 per cent of Swedish farmers belong to one or more associations, with an average of more than five memberships per farmer. In 1970 more than 80 per cent of Swedish farm produce was handled by these cooperative associations: 99 per cent of the milk delivered to dairies, 85 per cent of the beef and pork production, 36 per cent of the cured meat production, 75 per cent of the breadgrains and of the eggs. They operated 160 dairies, 36 slaughter houses and 33 meat curing and canning factories, 40 per cent of the flour mill capacity, 14 egg marketing centres and poultry processing plants, and distilleries and factories that accounted for the entire production of crude alcohol and starch from potatoes, etc. They also operated 477 rural credit societies/cooperative banks (with deposits of more than 4000 million Kr.) and 10 mortgage banks (with loans of 2600 million Kr.). They also supplied 70 per cent of the fertilizers and 65 per cent of the livestock feed concentrates purchased by Swedish farmers. Transactions between the associations and their members involved almost 15,000 million Kr. and their total employment exceeded 50,000 persons.

# IV Case Study of Sudan

## CONTENTS

# 15 Economic Implications of Recent Trends in Agricultural Production

The expansion of Sudanese agricultural production has historically been based on a strategy that placed major emphasis on expanding the area under cultivation in both irrigated and rain-fed areas. During the 1960s the area under cultivation increased by an average of 3.7 per cent per year and agricultural output increased by only 4.3 per cent per year (between 1960–4 and 1965–9 at constant 1969 prices), with crops increasing at a slightly higher rate than livestock. The index of agricultural production, on a per caput basis, varied between 122 in 1971, 110 in 1973 and 120 in 1974 (1961–5 = 100).

The Sudanese economy is dominated by agriculture through this sector's contributions to food consumption, economic growth, foreign trade, national income and employment. Agricultural performance necessarily conditions the basic features of the whole economic process, domestically as well as in relationships with the outside world. The national economy has accordingly been affected by the serious constraints under which agricultural production has been forthcoming. Recent trends in agricultural production have had important implications in terms of certain basic areas that deeply concern the economy (society) as a whole. These areas are food consumption, foreign trade and economic growth.

## FOOD CONSUMPTION

Available food supply provided an average of 2120 calories per caput per day in 1970. This is considered to be about 230 calories below estimated requirements. The index of per caput food production in the first half of the 1970s has remained within a range of 10 to 20 per cent above the 1961–5 average. Food consumption levels in the Sudan are not generally considered to be especially unsatisfactory when compared to the food situation in other developing countries and in view of the low per caput income. Nevertheless, food shortages occur in some areas as a result of shortfalls in local production which cannot be overcome by shipment of food from surplus areas because of transport difficulties.

Although the protein content of the diet is satisfactory, the diets of lower socio-economic groups in the country are considered deficient in animal protein, and a low intake of vitamins has been noted among the general population. To the extent that nutrition affects the state of health, it is significant to note that 19 per cent of children die during the first year of life and that the average life expectancy is only 40 years. To supplement local food consumption it has been necessary to import increasing quantities of foods and beverages, but the largest proportion of these imports has been consumed by the urban population. To overcome food shortages and to reduce food imports it is necessary to expand food production, including import substitutes. This would require in the first place a reduction of the present constraints that are impeding agricultural production expansion (see Chapter 16).

## FOREIGN TRADE

The composition of Sudanese exports has always been limited almost entirely to agricultural products. Cotton lint dominates exports and accounts for about 60 per cent of total export earnings. Exports of oilseeds (consisting mostly of cottonseeds, groundnuts and sesame), gum arabic and livestock account for 17, 8 and 5 per cent respectively. Other exports which together account for about 10 per cent of the total value include vegetable oils, oilcakes, hides and skins. This export mix reflects the heavy reliance on one major export commodity, cotton. With frequent fluctuations in cotton production as well as in world prices, the Sudan's foreign exchange earnings have been subject to great instability which has been reflected in the over-all performance of the economy. Moreover, with the exception of cottonseed oil and oilcakes, all exports are raw materials. The average annual rate of growth of total exports over the last decade averaged 3.3 per cent, while cotton exports grew at a higher rate of 5.8 per cent due to the increase in the volume of production and rising prices for Sudanese cotton.

The fact that the rate of increase of total exports approximated the rate of growth in agricultural production is to be emphasized. This means that the possibilities for expanding exports depend on increases in agricultural production. To the extent that expansion of production is impeded by various constraints, the expansion of exports is also impeded correspondingly. With the price boom of 1973/74, the value of exports increased by 15 per cent as compared with the previous year. As a proportion of GDP, exports maintained a relatively constant relation of 18 per cent during the period 1966–70. This proportion increased to an average of 21 per cent over the period 1972–4.

Over the last decade imports continued to increase, with quite pronounced fluctuations. Relative to GDP, imports have also shown a rising trend over the last ten years. They averaged 19 per cent of GDP over the period 1964–71, and with the rising prices of 1973/74, the ratio increased to about 25 per cent. Imports of consumer and intermediate goods rose during the last decade at an average annual rate of 3.5 per cent, and accounted for an increasing share of total imports, in contrast to imports of capital goods which declined at an average annual rate of 7.2 per cent.

TABLE 15.1 Percentage distribution of imports by main categories

| | *Annual average* | | *Growth rate* |
|---|---|---|---|
| | *1961/62-1965/66* | *1966/67-1970/71* | |
| Consumer goods | 45.2 | 50.5 | 4.0 |
| Intermediate goods | 34.4 | 36.2 | 2.8 |
| Capital goods | 20.4 | 13.3 | −7.2 |
| Total | 100 | 100 | 1.7 |

As imports of consumer goods expanded, imports of food, beverages and tobacco also rose to about 20 per cent of total imports. This ratio increased further to 22 per cent in 1972. The food import basket consists of wheat and wheat flour, sugar, tea, coffee, rice, dairy products, and canned fruits and vegetables.

Throughout the last decade imports have always exceeded exports. However, the trade deficit showed a tendency toward improvement as demonstrated by its different relative size at the beginning and at the end of the period under consideration. During 1961–5 the trade deficit represented about 10 per cent of the total value of trade. During the next five years it dropped to about 4 per cent. In 1972 there was a small surplus of £S 2 million. In 1973 the deficit rose to £S 22.7 million, but as a proportion of the total value of trade this deficit amounted to only 7.3 per cent. The trade deficit could have been much bigger with the present needs for interests of capital, intermediate and consumption goods, but it had been controlled by the government according to considerations of credit conditions and the balance of payments situation. Regarding the future, imports should expand more rapidly to increase the productive powers of the economy. However, exports as well as the production of import substitutes will have to be considerably expanded to do so and, in the present state of Sudanese development, it is only agriculture that can provide the needed expansion of both exports and import substitutes. This cannot be achieved without reducing the constraints handicapping the expansion of agricultural production.

## ECONOMIC GROWTH

As the Sudanese economy developed over the period 1966–70, the agricultural production growth rate of about 4.3 per cent was modest. Also modest was the growth of exports, GDP and agricultural GDP – all increasing at about 3.5 per cent. The increase of per caput income was very small and the atmosphere was one of relative economic stagnation in which investment and savings decreased to 14 per cent and 12 per cent respectively of GDP. When agricultural production decreased from 1971 to 1973 by about 3.3 per cent, the rate of growth of GDP went down to 3 and 3.5 per cent in 1971 and 1972, respectively. Investment went down from 15.1 per cent to 10.8 per cent of GDP, savings went down to 5.8 and 2.5 per cent of GDP in 1971 and 1972, respectively. The rate of growth in agricultural production thus conditioned the performance of the entire economy; when agricultutural production slackened or decreased, all the major economic parameters were adversely affected. National economic development in the Sudan must therefore be spearheaded by agricultural development because this is the major sector for generating income, providing exports, domestic food supplies and raw material for industry. There are, moreover, vast agricultural resources awaiting development, while industry is still a minor sector.

In addition, the possibilities for expanding agricultural exports depend on the expansion of agricultural production. As developments have clearly demonstrated, the rate of growth of agricultural exports almost equals the rate of growth of agricultural production. The possibility of expanding exports from present production is practically non-existent. Such expansion would have to be at the expense of a reduction in domestic food consumption. With the supply of calories per caput per day below estimated requirements, it is simply not possible to reduce food consumption in order to expand exports. Nor is it possible from a practical point of view to reduce imports of consumption and intermediate goods in order to save foreign exchange for use to expand imports of capital goods. The importation of consumption goods is necessary to maintain present modest consumption levels.

Increased investment from Arab regional sources in Sudanese agriculture so as to provide supplies for regional markets will lessen but not remove foreign exchange constraints on imports of capital equipment and other external inputs needed for agricultural expansion. Moreover, additional imports of capital goods which must be financed largely by expanded agricultural production and exports are only one requirement for expanding investments. Another important requirement is savings to cover that part of the cost of investment projects (in addition to imported capital goods) which must be financed domestically. In the Sudan, agricultural production is the primary source of these savings. In short, reduction of the various constraints on agricultural production is a priority prerequisite not only for agricultural development but also for national economic development.

# 16 Present Constraints on Agricultural Production

Constraints of an economic, technical, institutional and organizational nature are responsible for the relatively modest increase of total agricultural production and per caput production during recent years in the Sudan. These constraints combine to impede the expansion of both crops and livestock production.

## SUBSISTENCE AND QUALITY CONSTRAINTS

Agricultural development has historically been very uneven among the various regions in the Sudan. The developed areas are mostly concentrated and capital-intensive irrigation schemes which account for a large portion of the agricultural income and for the limited number of export crops on which the economy relies for its foreign exchange earnings. The bulk of the agricultural sector is, however, traditional and largely of a subsistence nature in the west and south. The inter-regional inequalities in development are manifested in the distribution of per caput incomes which, in the least developed regions, are about one-third of the level in the developed areas, and a quarter of per caput non-farm income. The concentration on large public irrigation schemes to the neglect of both the smallholders and rain-fed agriculture is putting a brake on the expansion of production. Unless the involvement of the smallholders is secured and their production oriented to the market, they will continue to produce just for their home use with very limited quantities going into exchange. In addition, the neglect of the large rain-fed savanna belt represents the loss of large potential that would not require heavy capital inputs or long gestation periods. The temptation exists for further development of the relatively well-developed area on the basis of mere convenience, short-term economic benefits or even political expediency. In any case, underdevelopment of the long-neglected regions continues and the transformation of the smallholders is not secured, with production potential lost and income inequalities perpetuated and increased.

ECONOMIC AND FARM INCOME CONSTRAINTS

In addition to the income inequalities within agriculture, there are also income inequalities between agriculture and the rest of the economy. With 80 per cent of the total population considered agricultural, the share in GDP of about 38 per cent underlines the crucial question of maldistribution of income. Per caput agricultural income in 1972/73 was about a quarter of per caput income in the non-agricultural sector. The reason for this disparity relates, in part, to the greater economic power and monopoly elements in the non-farm sector which favours the urban community.

The performance of agriculture is hampered by the characteristics of the economic mechanism and institutions that link it with other sectors of the economy. Marketing, credit and prices discourage production expansion and higher efficiency. With the exception of cotton, there is no organized marketing system for agricultural products. Cotton, having been accorded higher priority, is the subject of government participation in all stages from production to exports. The State Marketing Corporation was established in 1970 to act as a non-profit agent for all cotton exports and as such can render a satisfactory cotton marketing service. As to marketing other crops and livestock, farmers are adversely affected since their bargaining power is very limited. It is a common practice for producers to sell their crops in advance to secure merchant credit, and they are subjected to serious exploitation. Even when organized into cooperatives, they have not been successful because of financial and administrative difficulties. Furthermore, in the absence of organized markets and trading according to grades and standards, the quality of products has deteriorated and ethical standards have been frequently violated. Still more serious, production and exports are impeded by the inadequacies of present marketing practices and the tendency is clearly towards perpetuating subsistence production.

Closely related to marketing is the pricing of farm products. In the absence of an organized market or effective cooperatives, prices cannot be expected to perform their function as guides to production in the interests of both producers and consumers. The present cost-sharing arrangements in the irrigation schemes are rigid, and the allocation of resources among crops is decided regardless of price expectations. Cotton prices are set by the Sudan Cotton Marketing Corporation, and as this is a non-profit organization, producers are getting higher average prices for their crops. Other crops in the irrigation schemes are left for the tenants to grow and sell privately; but they cannot obtain remunerative prices and are frequently exploited – as are other smallholders in the traditional sector. The practice of selling in advance to secure credit also inflicts injustices upon farmers, and unless there is accessibility to institutional credit this practice will be perpetuated. The present market disorganization, merchant-dominated pricing, separation of prices from resource-use decisions and shortage of credit at reasonable rates, act among themselves to discourage production expansion, raising efficiency or improving quality. It is imperative to review periodically the present economic mechanisms and institutions to reduce their constraining effects on farm production and income, and to bring more consistency among them by taking into consideration their interdependence, conflicts and complementarity.

## CONSTRAINTS ON CROP PRODUCTION

One of the main reasons for the low level of productivity and the slow expansion of crop production is the very extensive way of using land. The land utilization ratio is low even in irrigated areas, while under rain-fed farming half the land is left fallow each year. Another important reason has been the decreasing yields of all crops, except cotton. The yields of sorghum, the basic food crop which occupies close to 45 per cent of the cultivated area, have been all too clearly decreasing; the 1.4 million tons crop produced from 4.2 million feddans in 1969/70 had been produced from 3.5 million feddans in 1960/61 (one feddan = 1.038 hectares). Yields have continued to decline further and in spite of the expansion of area between 1972 and 1974 production continued to decrease from 1.8 to 1.3 million tons. Expansion of area tended to bring land of a more marginal nature into production. A more important factor contributing to the decrease in average yields appears however to be the general lack of emphasis and incentives in policies and programmes for increasing yields of crops other than cotton.

With the modest expansion of the cultivated area, the noticeable rise in cotton yields and the decrease in yields of other crops, the rate of expansion of the agricultural production has been relatively low compared with the existing potential or the need for onput. The following tabulation shows the rates of growth of areas under basic crops and the rates of growth of production of these crops between the first and second five-year averages of the last decade. Subsequent developments do not seem to have altered the picture presented by these data.

Of major significance to the level of yields and slow expansion of production of the various crops (except cotton) is the quantity and quality of inputs used. Regarding the use of improved seeds, the situation is only satisfactory for cotton because a government corporation supplies farmers with such seeds. For other crops, it is the responsibility of farmers to undertake the arrangements for getting the seeds they need. This is true for sorghum, the most important foodgrain, as well as groundnuts and sesame which, after cotton, are most important in terms of product value and exports. Needless to say, any production development programme must be based on the use of improved seeds, which have to be supplied for all

TABLE 16.1 Average annual rates of growth in area and production of basic crops between 1960/61 – 1964/65 and 1965/66 – 1969/70

| | *Area* | *Production* |
|---|---|---|
| | (*percentage*) | |
| Cotton | 2.0 | 6.5 |
| Extra long and long staple | 1.4 | 5.6 |
| Medium staple | 16.2 | 34.0 |
| Short staple | −0.8 | −6.0 |
| Cereals | | |
| Sorghum | 2.0 | 0.2 |
| Millet | 5.0 | 1.0 |
| Wheat | 27.0 | 21.0 |
| Maize | −0.9 | −3.1 |
| Oilseeds | | |
| Cotton seed | 2.0 | 5.6 |
| Ground nuts | 6.7 | 5.5 |
| Sesame | 4.0 | −1.2 |
| Castor oil | 16.5 | 28.0 |
| Sugarcane | 12.5 | 32.0 |
| Others | 13.6 | — |
| Total | 3.7 | |

producers and for all basic crops in all areas in an organized and systematic fashion.

The use of fertilizer, another important input, is far below the economically desirable levels. All chemical fertilizers are imported since none are produced in the Sudan. The quantity imported amounts on the average to only half the estimated requirements. This means that some of the basic crops receive no fertilizer at all, and some are only partly fertilized, particularly since cotton fertilization is accorded higher priority. Modest average yields are therefore unavoidable, and in some areas yields must be very low even for cotton, which cannot get all its estimated fertilizer requirements with the quantities of fertilizer imported.

The plant protection service is, again, only satisfactory for cotton. For other crops it is inadequate in terms of both staff and finance. While about £S 5 per feddan was available for cotton protection in 1972/73, only £S 0.05 per feddan was available on the average for other crops. This amount is not sufficient to control locusts, grasshoppers, water hyacinth and other pests and diseases. It is estimated that losses in production amount to 10 per cent of the crops as a result of the insufficient coverage of non-cotton crops in addition to the perpetuation of pests and diseases. But cotton, in its turn, has been more subjected to pests on account of the loose application of the necessary preventive measures. This has resulted in the increase of the number of chemical sprayings from two in 1961 to five in 1971. With steep increases in the price of pesticides, cotton protection has become very costly, and it is necessary to apply strictly the preventive measures to reduce the need for chemical treatments.

## CONSTRAINTS ON LIVESTOCK PRODUCTION

The expansion of livestock production at an average rate of 3.6 per cent must be considered very slow in terms of both the existing livestock numbers and the export possibilities to neighbouring countries. Total livestock numbers are not precisely known, but are estimated at about 12.3 million cattle, 10.3m sheep, 7.2m goats and 2.5m camels. Nearly half the cattle population is concentrated in the western region where the Baggra cattle are raised under severe natural stress; the other half, concentrated in the southern region, is seriously subjected to insects and pests. In the eastern region there is a smaller number of cattle, and the Butana and Kenana regions have conditions of animal husbandry which are comparatively better than in other regions. Sheep and goats are more concentrated in the grassland of the savanna belt in the east, but are also raised in almost all other areas in varying numbers. Camels are mainly raised in the desert areas of the north.

The livestock sub-sector currently belongs to the traditional subsistence sector of the economy. It is predominantly nomadic and semi-nomadic husbandry with a low level of economic performance. It is practically separated from the rest of the agricultural sector and there is hardly any complementarity between crop and livestock production. Furthermore, cattle in the southern region are separated from the market, as cattle ownership is considered not as an economic asset but rather in terms of social status. The constraints under which livestock husbandry is practised and which do not permit the realization of the great potential of livestock can be summarized as follows.

1 The traditional, tribal, nomadic and transhunt systems of animal husbandry and the severe environmental conditions prevailing result in a low off-take rate, high death rate, late maturity of animals, marked seasonality of supply and low quality of meat.
2 Long distances between producing areas and the main consuming areas and export points, in which there are serious inadequacies in the system of stock routes, holding grounds, water points, forage reserves and lack of a developed all-weather road system or adequate rail transportation.

3 Deficiencies in the marketing of livestock and meat as market grades are not used; and with excessive marketing margins at the expense of producers, little incentive is left to undertake any improvement.
4 Deficiencies in policies for developing range conservation and improvement programmes, forage production, complementarity in the rotation between crops and livestock production, and improving and feeding practices.

The priority accorded to the livestock sub-sector is undoubtedly low, resulting in the shortage of experienced staff and funds to undertake all-round improvement of this important sub-sector. Livestock, which gives employment to about one-third of the total population, contributes only about 10 to 11 per cent of the GDP and about 6 per cent of total exports.

# 17 Strategy for Agricultural Development and Adjustment

There is no reason for the Sudan's rich agricultural resources to remain under-utilized and the country to continue poor. In view of the considerable resource potentials and the present state of technology, agricultural production in this country could achieve relatively high growth rates, provided effective local and external demand is expanded and appropriate development and production policies are applied.

The purpose of agricultural development in the Sudan is not simply that of increased production, but rather increased production based on an overall strategy geared to the fulfilment of several major objectives. These objectives relate to the generation of higher and better distributed incomes, the expansion and diversification of agricultural production and exports, self-sufficiency in food, integration of the traditional sector into the economy and achievement of more balanced regional development. These strategic objectives which have been adopted by the government constitute the primary guidelines for development and orientation of the Sudan's potential for increased crop and livestock production.

The particular options and related measures selected from among those available must clearly be conducive to achievement of these strategic objectives. Foremost among the policy measures that have been selected are those related to land and water resources development and their use in the expansion of crops and livestock production, and to the allocation of agricultural output as between domestic consumption and exports and the consequent implications for consumption patterns and the trade balance.

With regard to the development of land and water resources, investments in the past were concentrated on large-scale, capital-intensive irrigation schemes, principally for cotton with production strongly oriented for export. This trend is likely to continue to the extent that investments will be directed to fuller use of the existing under-utilized capacities in the export oriented schemes in the eastern region. Such investments would involve relatively small outlays of capital and result in high returns from the expansion of the exportable surplus in areas with good transportation facilities to Port Sudan. The continuation of the past trend of channelling the bulk of investments into these schemes could, as a policy option, be economically attractive and perhaps more convenient. But such an option would mean continued neglect of other regions, and the perpetuation and increase of regional imbalances and inequality in income distribution. It would also mean lower food self-sufficiency ratios and the continued strong orientation of production for foreign markets at the expense of building the national economy and expanding the domestic market.

The policy option regarding land and water resources development must clearly depart from the past trend if the strategic objectives are to be realized. The option selected must be aimed to push faster development in the long-neglected regions in the west and south and relatively more limited development in the irrigated east. It must therefore be designed to open, for modern agricultural production, both the rain-fed savanna to develop the west and the semi-equatorial region to develop the south.

In the development of the respective regions, the advantages of natural resources must be taken into account in the choice of the crops and livestock that need to be expanded in each location and in the selection of the magnitudes of the expansion needed to support achievement of the strategic objectives. Accordingly, crop production would be most important in the northern region, while both crop and livestock production would be important in the west. In the south, in addition to crop and livestock production, forestry and fishery would also play important roles.

### PRODUCTION FOR EXPORT EXPANSION AND FOR IMPORT SUBSTITUTION

In view of the large volume of foreign exchange required to finance import substitution in both agriculture and industry, emphasis on earning and saving foreign exchange cannot be reduced even in a framework of an inward-oriented development strategy. Consequently, the export expansion and import substitution objectives in terms of regional distribution of production would imply expanded production of sorghum, groundnuts and sesame for export from the rain-fed savanna, and the introduction and expansion of production of rice, sugar cane, coffee, tea, tobacco, kenaf and forestry products as import substitutes in the semi-tropical south. In guiding the choices among various production options in relation to these two objectives, the other broader objectives of diversification (of production and exports) and balanced regional development are, however, overriding. More specifically, choices among production possibilities have to be made not only within the context of each of the objectives, but also in relation to a balance between them.

Livestock development, in terms of export expansion and diversification, should also play an important role, as its development would contribute directly to balanced regional development, better distribution of income and complementarity within and among regions.

In the course of economic development, a choice has to be made regarding the possibly conflicting ends of expanding exports and of expanding domestic food consumption. The implications of such a choice depend clearly on the rate of growth of agricultural production, since supplies for both exports and for increasing domestic consumption must come from this production. With a low rate of growth, for example, a rate not much higher than the rate of population growth, the expansion of one of the uses of agricultural production can only be achieved at the expense of restricting or even reducing the other. With an envisaged growth rate of agricultural production as detailed later, it would be possible to expand both exports and domestic food consumption. The possibility of expanding both uses reflects the availability of rich agricultural resources in the Sudan and the possibility of their effective development.

The option chosen regarding the growth rate of agriculture (and consequently all the basic parameters of a macroeconomic growth variant) is of particular importance for two reasons. The first relates to the growth rate of agriculture and the economy as a whole which should be as high as can be practically achieved in order to speed up development. The second relates to the importance of agriculture in the Sudanese economy and the consequent dependence for achievement, not only of all the strategic objectives of agricultural development but also of all the basic objectives for the non-farm sector, on the realization of the rate chosen for the growth of agricultural production. The choice of a specific rate that would speed up development depends on complex technical, economic, institutional and organizational factors which must be taken into consideration in selecting this rate on the basis of both scientific procedures and judgement. Most clearly, the ceiling on the demands that can be placed on the agricultural sector is set by the highest possible rate of agricultural growth.

## THE AGRICULTURAL GROWTH RATE

In the Perspective Study of Agricultural Development in the Democratic Republic of the Sudan undertaken in 1973/74 by FAO, two variants of GDP growth rate supported by their relevant parameters were formulated for the period to 1985. The first variant assumes a quasi continuation of past trends. The second reflects a high growth path which corresponds with the growth targets of the government's five-year plan 1970–5. The following analysis focuses on the high growth variant of the FAO perspective study with the objective of clarifying its perspectives and requirements for achieving its goals.

The gross value of agricultural production in the growth variant under consideration is postulated to grow at an average annual rate of 5.4 per cent during the period 1970–85, with crops increasing at an average annual rate of 5.3 per cent and livestock at 5.8 per cent. These rates of growth would require an increase in the ratio of inputs to the gross value of agricultural production from 19 per cent to 26 per cent. The increase in the use of inputs would amount to 8 per cent per annum. With this increase in the use of inputs, the average annual rate of growth in agricultural GDP would be 4.7 per cent. These growth rates would increase agricultural income per caput from £S 19 in 1970 to £S 27 in 1985 (at constant prices). Agricultural investments during the period would amount to 11 per cent of agricultural GDP as compared with the envisaged over-all economy's investment coefficient of 19 per cent. With regard to external trade, agriculture would continue to provide all the supplies of export products, with an estimated average annual growth rate of 6 per cent. Agricultural imports would, however, decrease at an average rate of 3.9 per cent as a result of the very substantial expansion of import substitution.

For the economy as a whole, it is postulated that the GDP would grow at an average annual rate of 5.5 per cent. The growth rate would require the expansion of gross investment at an annual average rate of 7.4 per cent, raising the ratio of investment to GDP from 14 to 19 per cent by 1985. As a result the savings gap would widen and reliance on external resources would increase considerably. With the ratio of savings to GDP estimated to increase only from 12.5 per cent in 1970 to 13 per cent in 1985, the ratio of capital inflow to GDP would have to increase from 1.5 per cent to 6 per cent. Imports would increase at a higher average annual rate of 7.3 per cent in view of the vital role of capital imports in the accelerated investment effort.

During the last three years of the fifteen years horizon, the shift towards acceleration of inward-looking development would be emphasized. This orientation is needed to reduce the excessive reliance on exports and foreign exchange, to broaden the base of production, to integrate the traditional sector into the economy and to step up the development of manufacturing. Fulfilment of these objectives would move the economy towards increased self-sufficiency in food and agricultural commodities and to yielding higher and better distributed incomes.

The role of agriculture under this high growth variant is strengthened, and the significance of increasing agricultural exports is further stressed. The sector, however, has to meet the demand not only for more foreign exchange earnings, but also for saving foreign exchange by import substitution. Higher domestic demand for food will also necessitate a stepped up production of all food commodities and particularly the import substituting items. The sector is also envisaged as contributing to balanced regional development with particular emphasis on the smallholder and the livestock sub-sector. In addition, acceleration of the development of manufacturing would require that agriculture responds adequately in the provision of raw materials and serves as a market for manufactured goods.

## INPUTS AND INVESTMENTS

Modernization of crop and livestock production to achieve the stated objectives of agricultural development would require an increasing share of inputs and investments in the production process. The establishment of modern schemes in the rain-fed areas calls for relatively substantial investments in mechanization, drinking water facilities and transportation. For the annual and permanent crops which would be promoted in the rain-fed areas as import substitutes in the south, high investments and inputs are visualized in addition to the investments for related processing facilities.

Crop intensification in the irrigated areas requires more intensive use of such inputs as better-quality seeds of improved varieties, fertilizers, pesticides, mechanization and irrigation water. In addition, the expanded processing of farm output would require relatively substantial investments, particularly in sugar processing facilities.

The development of livestock production would also require increased inputs and investments. The improvement of traditional production would call for increased veterinary services and expansion of watering points, while modern production would depend on large increases of purchased feed. In addition, considerable investments would be needed in breeding ranches, holding grounds, feed lots, slaughtering houses and improved marketing facilities.

The annual cost of inputs needed for the envisaged development of agricultural production is estimated to increase from £S 49 million in 1970 to £S 145 million by 1985, with their ratio to gross value of agricultural production increasing from 19 to 26 per cent. Crop production would require about 75 per cent of the total inputs to cover irrigation, mechanization and fertilizers. Among the additional inputs needed for livestock production, the cost for providing drinking water would be the dominant item.

Over the period 1970–85, average annual investment requirements at the farm production level were estimated at £S 35 million, one-third of which would have to be allocated for livestock development. Mechanization, storage and the establishment of plantations would account for the remainder. Additional investments, to the amount of £S 21 million annually, would be needed for both transportation and agricultural processing facilities.

Total investments for production over the 1970–85 period would amount to £S 528 million and reflect an incremental capital–output ratio of 2.5. About 60 per cent of these investments would be in foreign exchange.[1] Their ratio to agricultural GDP would be 11 per cent over the period 1970–85 as compared to the envisaged over-all economy's investment coefficient of 19 per cent.

### INSTITUTIONAL SUPPORT AND ECONOMIC INCENTIVES

The objectives of agricultural development cannot be achieved by the mere provision of material investments and inputs. The whole agricultural institutional and economic framework must also undergo fundamental changes to ensure a more effective contribution of farm people to this production process. Without the mobilization and development of farm people and direction of their activities towards production, it would not be possible to achieve the envisaged objectives of agricultural development.

Foremost in the development process is the availability of skilled manpower that is properly educated and trained. Management ability, in particular, is necessary for the successful performance of large schemes and institutions. For the small farmers, a strengthened and expanded extension service is needed to support the modernization of the traditional sector. More intensified extension would be needed in the south, where new plantations and new techniques would be introduced. Agricultural corporations in the mechanized schemes should also provide effective extension service to their field workers in all aspects of production in the new environment.

Research on crops and livestock should be strengthened and basically oriented towards finding solutions to the various problems associated with the introduction of new crops, extension of crops to new ecological regions and improvement of livestock. Priorities should be clearly established for solving the problems posed by the more effective utilization of arable land in the tropical south and arid and semi-arid savanna, as well as the development of livestock potential and the exploitation of fishery resources. Due emphasis must be placed on strengthening and reorganizing the present agricultural research institutions and coordinating multi-disciplinary research activities required to solve such priority problems. In the socio-economic field, research is particularly needed with regard to programmes involving nomads, their settlement and their attitudes towards livestock in order to participate more fully in production. The vexing problems of nomadism would require the cooperation of several ministries and research institutions. It should not be discouraging that agricultural cooperatives have not so far been very successful. They should be strengthened through training for both staff and members to perform basic services, particularly in the distribution of inputs and marketing of products. Their development is essential for the small farmer, particularly in the traditional sector where their marketing of short-staple cotton, oilseeds, gum and livestock would support the development of that sector.

Essential to the envisaged development of agriculture is the expansion of institutional credit. The present system of institutional credit is inadequate and basically inaccessible to small farmers. As such they are forced to rely on the 'Sheil' system under which they receive advances in cash and kind from private merchants against promises to deliver crops after harvest. This system results in the payment of very high rates of interest which impedes the expansion of production. An adequate expansion of institutional credit is needed since credit needs would increase with the expanded use of inputs. It is not a simple matter to provide the finances needed for such credit expansion. Moreover, there are other limitations which relate to the lack of secure markets for agricultural products, to price fluctuations and inadequate transportation. These limitations must be gradually reduced in order to ensure a sufficient volume of credit with reasonable interest rates and simplified procedures to support development of production and improvement of the system of distribution.

Crop prices have an immediate bearing on farmers' incentives to produce. Producer prices are generally low and tend to impede expansion of production. The formulation of an adequate farm price policy geared to fulfilment of the envisaged production targets for major products is necessary and should be combined with export taxes that make Sudanese exports competitive in world markets. Import duties on agricultural inputs should be reduced where possible to lower prices and thereby to stimulate increased use of these. Duties levied against some import commodities that could be grown domestically are considered by the government to be such significant sources of revenue that domestic production of the commodities to substitute for imports is discouraged. A workable adjustment of agricultural import duties to take account of both short-term and long-term costs and benefits is needed. Closely related to agricultural cooperatives, price policy and institutional credit is the marketing system with its excessive margins, low standards of handling methods and waste. These pose major impediments to the expansion of production of several agricultural products. Improvement of the marketing system requires that, in addition to the involvement of cooperatives and credit institutions in marketing, the government should also get involved by establishing marketing corporations to reduce undue margins, provide price leadership in various commodities and play a vital role in meeting export targets.

## DEVELOPMENT OF CROP PRODUCTION

The expansion of crop production during the period 1970–85 rests basically on expansion of the harvested area and on improvement in yields. Because of the availability of land, the Sudan can obtain much of the future growth in agricultural production from area expansion. It has already been mentioned that the strategic objectives of agricultural development require a departure from the past tendency to channel the bulk of investments into major irrigation schemes and to provide for a faster expansion in the rain-fed areas in the savanna and the south. It is estimated that during this period an increase of 50 per cent in the irrigated harvested area and an increase of 80 per cent in the rain-fed area would provide the harvested area needed to achieve the targets for increased crop production. The projected expansion in both areas would expand the total cultivated area by 15.6 million hectares and the harvested area by 5.6m hectares. The total cultivated area is estimated to expand from 26m hectares in 1970 to 41.6m hectares in 1985.

In the irrigated area, which is estimated to expand by 1.2m hectares from 1970 to 1985, emphasis is placed on intensification and on assigning a larger place in the rotation to high value crops such as medium-long staple cotton, fruits, vegetables and sugar cane. Cropping intensity is estimated to increase from 73 per cent in 1970 to 80 per cent in 1985. In contrast the larger expansion in rain-fed areas is associated with only minor increases in yields and slight improvement in rotation, with emphasis placed on a considerable expansion of the area under sorghum, groundnuts and sesame seeds. As a result of changes envisaged in area, intensity and yields, it is estimated that the share of rain-fed areas in gross crop value would rise from 42 per cent in 1970 to 52 per cent in 1985. At the same time, the share of irrigated areas in total harvested area would decline from 21 per cent to 18 per cent.

Of the 8.6m hectares increase envisaged in harvested area between 1970 and 1985, 7.5m hectares would be allocated to the production of traditional basic crops such as cereals, oilseeds and seed cotton. Significantly, sorghum and millet alone would account for fully half the total increase in harvested areas. Oilseeds would come second with an increase in harvested area of 1.8m hectares and seed cotton would follow with 0.8m hectares.

Although changes in the land-use pattern would be quite limited as a result of allocating the additional area among basic crops, the orientation of crop production to the basic objectives of domestic food demand, export expansion and import substitution is more visible in terms of harvested area and crop value, as shown in Table 17.1.

TABLE 17.1 Relative orientation of crop production

| | *Harvested area* | | *Crop value* | |
|---|---|---|---|---|
| | *1970* | *1985* | *1970* | *1985* |
| | *(percentage)* | | | |
| Domestic food demand | 77.7 | 71.5 | 52.3 | 63.3 |
| Export expansion | 19.4 | 23.7 | 44.1 | 27.5 |
| Import substitution | 2.9 | 4.8 | 3.6 | 9.2 |
| Total | 100.0 | 100.0 | 100.0 | 100.0 |

It is readily clear that in 1970 crop production reflected a significantly outward-looking orientation, with exports accounting for 44 per cent of the value of crop production. This high percentage was contributed by less than one-fifth of the harvested area. By 1985 production is estimated to reflect a more inward-looking orientation, with the share of import substitution rising to almost 10 per cent of total crop value. In addition, the share of exports in the value of crop production would drop to 27.5 per cent, largely because of the higher share for domestic demand for food.

In terms of contribution to the growth of agricultural production, the increase in the value of crops (at 1970 prices) between 1970 and 1985 is estimated at £S 227 million, an average annual rate of increase of 5.3 per cent which compares to an average annual rate of increase of total agricultural production of 5.4 per cent. This increase in the value of crops would amount to 66.4 per cent of the increase in the total value of agricultural production. The share of crops in total agricultural production is, however, estimated to decrease from 68.2 per cent in 1970 to 67.2 per cent by 1985, with the share of cotton decreasing from 29.2 to 24.4 per cent and the share of cereals increasing from 12.4 to 15 per cent.

TABLE 17.2 Perspectives of crop increases by 1985

| | *Production* | | *Value* | | *Area* | |
|---|---|---|---|---|---|---|
| | *'000 tons* | *Average annual growth rate* | *£S (million)* | *% of total increase* | *Feddans*[a] *(million)* | *% of total increase* |
| Cereals | 2871 | 6.7 | 58.3 | 25.7 | 4.7 | 56.4 |
| Oilseeds | 801 | 6.1 | 25.8 | 11.4 | 1.7 | 20.2 |
| Starchy roots | 1395 | 4.2 | 11.0 | 4.8 | 0.6 | 7.0 |
| Seed cotton | 574 | 4.2 | 70.0 | 30.8 | 0.8 | 9.6 |
| Fruit and vegetables | 1000 | 5.0 | 38.0 | 17.0 | 0.2 | 2.4 |
| Others | | | 24.0 | 10.3 | 0.4 | 4.4 |
| Total | | | 227.1 | 100.0 | 8.4 | 100.0 |

[a] One feddan = 1.038 hectares.

Of the total increase in the value of crops the traditional commodities (cotton, cereals and oilseeds) would contribute 68 per cent. If fruits and vegetables are added, their total contribution would amount to 85 per cent. In physical terms, cereal and oilseed production are estimated to increase at an average annual rate of 6.7 and 6.1 per cent respectively, fruits and vegetables at 5 per cent and cotton at 4.2 per cent.

DEVELOPMENT OF LIVESTOCK PRODUCTION

The present numbers of livestock and the vast grazing areas in the Sudan can be developed to contribute significantly to the growth of agricultural production, the expansion and diversification of exports and the improvement of domestic nutritional standards. Moreover, livestock development would contribute to balanced regional development since it would raise the generally low incomes that prevail in the livestock producing regions.

In terms of contribution to the growth of agricultural production, the increase in the value of livestock production from 1970 to 1985 is estimated at £S 78.4 million (in 1970 prices), an average annual rate of increase of 5.8 per cent. This increase amounts to 23 per cent of the estimated increase in the total value of agricultural production. As a result of this development, the share of livestock in total agricultural production is estimated to amount to 22 per cent in 1985, as compared to 21 per cent in 1970. The entire increase in the share of livestock would be due to the envisaged expansion of the cattle and sheep industries, which would contribute 16.6 per cent of total agricultural production as compared with 14.8 per cent in 1970.

Within the livestock industry, cattle expansion would be dominant, with an average annual growth rate of 6.6 per cent; sheep would follow with 5 per cent, and other livestock with 4.4 per cent. Of the £S 78.4 million increase in the annual value of livestock production as between 1970 and 1985, cattle would contribute 45.4 per cent, sheep 34 per cent and other livestock 20.6 per cent. The estimated contribution of livestock to agricultural production development is summarized in Table 17.3.

TABLE 17.3 Estimated contribution of livestock to agricultural production

| | *Average annual growth rate* | *Share in value of total agricultural production* | | *Increase in value* | |
|---|---|---|---|---|---|
| | *1970–85* | *1970* | *1985* | *1970* | *to 1985* |
| | | *(percentage)* | | *(£S m)* | *(% of total)* |
| Cattle | 6.6 | 7.8 | 9.2 | 35.6 | 45.4 |
| Sheep | 5.0 | 7.0 | 7.4 | 36.6 | 34.0 |
| Others | 4.4 | 6.2 | 5.5 | 16.2 | 20.6 |
| Total | 5.8 | 21.0 | 22.1 | 78.4 | 100.0 |

Regarding expansion of livestock products, the dominant role of cattle would be further emphasized in the expansion of meat and milk. The share of beef and veal in total meat production is estimated at 58 per cent in 1985 as compared to 50 per cent in 1970; of a total expected increase of 497,000 tons in meat production, the share of beef and veal would amount to 62 per cent, and the share of mutton, lamb and goat to 19.4 per cent. As to total milk production, the share of cattle milk is estimated at 74 per cent in 1985 as compared to 65 per cent in 1970; of an estimated increase in total milk production of 1.6 million tons, the share of cow milk would amount to 83 per cent.

Poultry production, which amounted to 11,000 tons in 1970, is expected to double by 1985 but would still account for a very small part of total meat production. Egg production is estimated to increase from 17,000 tons to 47,000 tons during the same period.

The improved performance of the livestock sub-sector would aim mainly at developing a vigorous export-oriented industry with emphasis on cattle, sheep, and hides and skins. For this purpose supporting programmes are needed for commercial production and for building up the necessary structure for marketing, processing and export sales.

With respect to meat cattle production, the development policy would have to aim at slowing down the herd size growth rate to 2 per cent per year and at achieving considerable increases in productivity by introduction of major changes in the management of herds and their structures to eliminate non-productive animals. The control of herd size would leave a large proportion of the calf crop for meat production and help to stop the further deterioration of grazing resources. Further production development envisages the integration of crop and livestock production in the east by the establishment of modern and intensive lines of livestock production. Of particular significance in this production are the fattening operations based on young cattle raised in the traditional semi-nomadic sub-sector, and the development of commercial dairy herds of an adequate number to alleviate the milk shortage in the east.

From the feeding point of view, the attainment of livestock development objectives requires the improvement of natural grasslands, expansion of forage production in both irrigated and rain-fed lands, and the development of an animal feed industry. The future improvement of the traditional sub-sector rests basically on the improvement of natural grasslands in order to raise their carrying capacity by 1985 to 20 million animal units (80 per cent of their potential carrying capacity) as compared with the 15 million animal units presently supported. Major investments, particularly for the provision of drinking water, are needed. About 15,000 water points at a cost of £S 153 million would have to be developed by 1985 in the savanna belt in order to achieve the objective of improving and expanding the grazing land.

Fodder production has to be expanded in the irrigated areas to support the fattening and dairying activities in the east. The total green fodder required for the two activities would amount to 5 million tons by 1985. It could be produced from 148,000 hectares and could support 200,000 milking cows and the fattening of about one million animal units. The production of fodder would not however be limited to irrigated areas as there is great scope for the integration of crop and livestock production on rain-fed lands, particularly in the mechanized sorghum schemes in the east and the groundnut schemes in the west.

The development of dairy and poultry production would require the development of an animal feed industry. Concentrated feeds amounting to 100,000 tons for the dairy industry and 96,000 tons for poultry would have to be processed. In addition, supplementary protein and mineral concentrates would be needed for calves in selected areas in the west and south.

## NOTES

1 As noted in the introductory chapter, a much greater investment programme based on funds from Arab oil countries is now being undertaken in the Sudan. This represents an accentuation of the policy option of increased production of food for export in the course of overall economic development.

# 18 Implications of the Strategy for Agricultural Development

Because of the predominant role of agriculture in the Sudanese economy, the rate at which agricultural production is expanded has far-reaching implications for the national economy and for the well-being of the entire population. Successful application of the strategy of change in agriculture as set out in the previous section would have distinctly favourable implications, especially in terms of export expansion, import substitution and domestic food consumption levels.

## IMPLICATIONS FOR EXPORT EXPANSION

The growth of agricultural production at an average annual rate of 5.4 per cent during the period 1970–85 would permit the growth of the volume of agricultural exports at an average annual rate of 6 per cent during the same period. Such a growth rate would indeed be necessary in order to make possible the realization of the postulated growth rate of the GDP. The growth rate of exports during the periods 1970–5, 1975–80 and 1980–5 would amount to 5.0, 6.6 and 6.5 per cent respectively. This development reflects the emphasis on export expansion and inward-looking orientation of the development starting by 1980.

The estimated growth of export earnings from £S 102.3 million in 1970 to £S 240.7 million in 1985 (at 1970 prices) would be attributable to: (1) the increase in exports of major agricultural commodities (sorghum, cotton, oilseeds, livestock and livestock products); (2) a larger share of these exports in the form of processed products (cotton yarn, meat, oils, oilcakes and processed hides and skins); (3) emphasis on exports of high value commodities such as livestock, meat and oils; (4) introduction of some new export commodities such as fruits, vegetables and certain oilseeds.

About £S 129 million or 93 per cent of the estimated increase in export earnings would be attributable to traditional Sudanese export commodities, to larger volumes of these, and to additional pre-export processing of them. The continued reliance on traditional commodities for export expansion reflects the Sudan's comparative advantage in the production of these and its considerable potential for rapid expansion of this production. It also reflects the absorptive capacity of foreign markets and the level of expected returns.

Considerable differences in the rates of export expansion of the various products would result in the realization of the second objective, namely, the diversification of exports. The all too dominant role of cotton as the major export product would decline considerably. Lint cotton, which accounted for 64 per cent of total exports in 1970, is estimated to account for only 30 per cent in 1985; if yarn is added, cotton and cotton products would account for 43 per cent of all exports. Although the relative importance of cotton exports would decline, their absolute value would increase by £S 37 million, or 27 per cent of the total increase in the value of exports. This decline in the share of cotton would contribute significantly towards reducing the vulnerability of the economy that results from excessive reliance on one major export commodity.

The decline in the share of cotton would be offset by an appreciable rise in the share of oilseeds, oils and oilcakes, livestock and livestock products, and sorghum. Exports of these products would increase by £S 92 million, or 66 per cent of the total increase in the value of exports, and their share in the value of total exports would increase from 27 per cent in 1970 to 50 per cent by 1985. The estimated changes in exports are summarized in Table 18.1.

TABLE 18.1 Perspectives of agricultural export expansion and diversification

| | *Value (at 1970 prices)* | | | | *Increase in Value* | |
|---|---|---|---|---|---|---|
| | *1970* | | *1985* | | *1970 to 1985* | |
| | *£S million* | *percent* | *£S million* | *percent* | *£S million* | *percent* |
| Cotton and cotton products | 65 | 64 | 102.3 | 43 | 37 | 27 |
| Sorghum | 0.1 | [a] | 20.7 | 9 | 20.6 | 15 |
| Livestock and livestock products | 6.1 | 6 | 38.4 | 15 | 32.2 | 23 |
| Oilseed, oils and cake | 20.9 | 20 | 60.2 | 25 | 39 | 28 |
| (Sub-total) | (92.1) | (90) | (221.6) | (92) | (128.8) | (93) |
| Others | 10.2 | 10 | 19.1 | 8 | 8.9 | 7 |
| Total | 102.3 | 100 | 240.7 | 100 | 137.7 | 100 |

[a] less than 0.5 per cent.

With regard to the locational aspects of export expansion, these relate not only to the suitability of ecological conditions and the desirability of a better regional distribution, but also to the achievement of a more rational pattern of production in terms of the concentration of production for export in areas with ready access to the transportation system and close to the seaport. This concentration is necessary because the country is so poorly equipped with infrastructure that it would hardly be feasible to assign the envisaged increases in production of export-oriented crops to regions distant from the sea outlet. Thus, there can be only a limited shift of emphasis from further development of the modern, export-oriented schemes in the regions closer to the seaport to the development of new export-oriented schemes in distant regions in the south and west. Production of major commodities for export will need to continue to be concentrated in large schemes in the eastern region within the time horizon of this study.

Attainment of the export estimates for major commodities is judged feasible in terms of international market situations and prospects. It appears that the long-term world market prospects for increased cotton exports are improving due to the effects of higher oil prices on costs of synthetic fibres. Furthermore, the Sudan is involved in a number of bilateral trade agreements and no major problems should be anticipated in marketing its increased cotton exports and maintaining its share in the world market. There exists also a reasonable market potential for processed cotton exports, in particular for yarn.

The outlook in the world market for other major Sudanese products is also very favourable, particularly after the re-opening of the Suez Canal. Major increases are expected in the export earning of oilseeds, oils, sorghum, cattle and meat. On account of the strength of the international market for feedgrain, large increases in sorghum exports could be absorbed at remunerative prices. This is also true for oilseeds, oils and oilcake. Foreign markets for meat will certainly be expanding in the neighbouring oil-exporting countries. The European market could also be opened for Sudanese meat if veterinary services, upgrading and marketing arrangements were to be effectively improved. In general, the long-term external market prospects are such that the increase envisaged for Sudanese export volumes should be quite feasible, provided market access is actually available.

## IMPLICATIONS FOR IMPORT SUBSTITUTION

Although the Sudan's imports of foods and beverages have been steadily rising, there are promising potentials for expanding domestic production of these commodities. Higher self-sufficiency ratios in such deficit food products as wheat, rice, sugar, fruits and milk, and in such non-food products as tobacco and jute have been emphasized. National self-sufficiency in food is one of the strategic objectives of the Sudan's agricultural development policy.

The selection of import substitutes to be introduced and expanded would be guided by the suitability of ecological conditions (which largely determine relative costs of production amongst different locations), the contribution to balanced regional development and to transformation of the traditional sector, the magnitude of foreign exchange earnings, the level of investment requirements and the impact on agricultural processing and inter-sectoral linkages.

A major policy element in the above criteria relates to locational aspects. Contrary to the concentration of expanding production for increasing exports, dispersion would characterize the expansion of production for import substitution, with locational priority given to the least developed regions in order to achieve more balanced regional development. The expansion of tea, coffee, kenaf, rice and to some extent sugar production would be in the south on the basis of ecological and economic considerations. This would contribute to the badly needed development in that region. The west would benefit from expanded production of oilseeds, livestock and tobacco which would replace fully the imports of oils and fats and partly the imports of tobacco. In addition to increasing national self-sufficiency ratios and contributing to balanced regional development, expanded production of import substitutes would clearly strengthen the complementarity between the various regions and foster specialization in agricultural production.

The range of choices within the import substitution context is largely dictated by the composition of agricultural commodity imports and by the levels of demand projected for 1985. The latter would impose a ceiling on the volume of production needed to replace imports. If production were to surpass this volume, there would be a surplus for export or for additional expansion of domestic consumption levels. Further limitations to the range of choices relate to ecological factors, economic considerations and other constraints relating to organization, know-how and manpower requirements, so that it would be necessary to phase the replacement of imports by domestic production.

With expanded production of import substitutes, the complete replacement of imports of sugar, rice, pulses, fruit and vegetables is envisaged. Imports of other commodities such as wheat, tea and coffee would, however, be only partly replaced. With the exception of tea which would continue to be imported in increasing quantities, all other imports would decline or would be completely replaced. In the case of sugar and rice, production would cover domestic consumption requirements and in addition provide a surplus of sugar for export and of rice for partial replacement of wheat imports.

It is estimated that as a result of import substitution, the agricultural import bill would be reduced from £S 24.5 million in 1970 to £S 13.8 million by 1985, an average annual reduction of 3.9 per cent. On the basis of the trend value of agricultural imports extended to 1985 (£S 45.4m), the estimated saving would, however, be much greater (£S 31.6m).

Although self-sufficiency in basic foods is an emphasized policy objective, its realization with regard to wheat is not possible due to ecological considerations and constraints on area and yield. Consequently, the estimated wheat requirements of 837,000 tons in 1985 would only be partially met by expansion of domestic production in the eastern region to about 420,000 tons. The gap in wheat supply would be met by continued imports and by the substitution of cassava and rice. The potential for expansion of rice production is quite substantial in the southern region with its water availability and semi-flooding conditions. It is envisaged that from 1980 self-sufficiency in rice could be achieved, including the quantity needed to substitute wheat.

With regard to sugar, self-sufficiency and a surplus for export are expected from 1980. This should be quite significant since total sugar consumption has recently been increasing at an average annual rate of 7 per cent and production presently accounts for only 37 per cent of domestic consumption. Expansion of sugar production would be mainly in the eastern region. By 1985, self-sufficiency ratios in coffee and tea are estimated to amount to 64 and 20 per cent respectively. Considerable production adjustments have to be undertaken before such ratios could be achieved. On account of the expected slow expansion of tea production and the expansion of domestic consumption at an average annual rate of 4.5 per cent, imports of tea would continue to increase and by 1985 would account for 50 per cent of the total value of agricultural commodity

TABLE 18.2 Perspectives of agricultural import substitution

| | *Deficits* | | *Value of imports* | | *Self-sufficiency ratio* | |
|---|---|---|---|---|---|---|
| | *1970* | *1985* | *1970* | *1985* | *1970* | *1985* |
| | *('000 tons)* | | *(£S million)* | | *(percentage)* | |
| Wheat and flour | 226 | 318 | 5.4 | 3.4 | 29 | 40 |
| Tea | 19 | 27 | 5 | 6.9 | 0 | 23 |
| Coffee | 14 | 9 | 1.9 | 1.2 | 0 | 64 |
| Sugar | 1333 | +527 | 5.4 | 0 | 40 | 113 |
| Rice | 16 | +33 | 0.7 | 0 | 16 | 165 |
| Jute | 28 | 12 | 3 | 1.3 | 0 | 69 |
| Tobacco | 0.76 | 0.7 | 1.4 | 1 | 43 | 80 |
| Other | | | 1.7 | 0 | | |
| Total | | | 24.5 | 13.8 | | |

imports. Coffee production could be more easily expanded, and it is estimated that by 1985 coffee imports would be reduced to 9000 tons.

With regard to imports of jute bags, it is estimated that by reducing domestic requirements for these, through bulk handling of the increasing agricultural output and by increasing the production of kenaf to 27,000 tons, a self-sufficiency ratio of around 70 per cent could be achieved by 1985. Table 18.2 summarizes the 1970 commodity situations deficits and those projected for 1985.

## IMPLICATIONS FOR FOOD CONSUMPTION

The emphasis in the strategy for agricultural development in the Sudan on export expansion raises questions regarding its effects on domestic food consumption levels. More specifically, would the export expansion be achieved partly at the expense of domestic food consumption levels? Or can the increase in production be sufficient to permit the expansion of both exports and domestic food consumption? These questions are of basic importance in the context of agricultural adjustment.

The extent of the total increase in food availabilities for domestic consumption expected between 1970 and 1985 amounts to 74 per cent, with livestock products and fruits and vegetables reflecting the highest increase, as shown in Table 18.3.

TABLE 18.3 Food availability of major food commodity groups

| | *1970* ('000 tons) | *1985* | *Increase* (*percentage*) |
|---|---|---|---|
| Cereals | 1707 | 2917 | 71 |
| Starchy roots | 1469 | 2382 | 62 |
| Sugar products | 213 | 404 | 90 |
| Fruit and vegetables | 915 | 1773 | 94 |
| Livestock products | 2072 | 4018 | 94 |

With population growth taken into account, the increases in food commodities would permit increases in per caput consumption of major foods, as shown in Table 18.4.

In contrast with the increased supplies of higher value foods, supplies of starchy roots, millet and sorghum reflect the lowest increases. This is indicative of the gradual substitution of cereals for cassava and yams in the south, and of wheat for millet and sorghum in the north.

As a result of the quantitative and qualitative change in per caput food availability between 1970 and 1985 diets would improve – as measured in

TABLE 18.4 Per caput food availability of selected commodities

| | *1970* | *1985* | *Increase* |
|---|---|---|---|
| | *(kilogrammes)* | | *(percentage)* |
| Wheat | 16.6 | 23.3 | 40 |
| Fruit and vegetables | 59.7 | 74.4 | 25 |
| Sugar | 13.9 | 16.9 | 22 |
| Meat | 22.5 | 29.4 | 31 |

TABLE 18.5 Per caput daily intake of calories, proteins and fats

| | *1970* | *1985* | *Increase* |
|---|---|---|---|
| | *(kilogrammes)* | | *(percentage)* |
| Total calories | 2198 | 2511 | 14 |
| Total protein | 65 | 76 | 17 |
| Total fats | 60 | 72 | 19 |

terms of per caput daily intake of calories, proteins and fats. The extent of this improvement is shown in Table 18.5.

These increases in domestic consumption, which would bring about a significant improvement in nutritional conditions, could be achieved at the same time as agricultural exports more than doubled, with the largest expansion taking place in food and feed commodities. Without such an expansion of exports the agricultural potential of the country would not be fully tapped, because Sudanese domestic demand alone would provide insufficient markets for the volume of supplies which, it is judged, could be produced. Over-all economic development would be seriously hampered by the slower growth in export earnings. In brief, an examination of Sudanese conditions and prospects demonstrate that when potentials exist for a sustained expansion of agricultural production in a developing country, increased food exports and increased per caput domestic food consumption are likely to constitute a complementary objective rather than exclusive policy alternatives.

# Appendix: FAO Reports on International Agricultural Adjustment

The following studies have been completed by FAO in its programme of work on international agricultural adjustment over the period 1972–5:

1 International Agricultural Adjustment. C 73/15, September 1973. (Director-General's report to the 17th Session of the FAO Conference, November 1973.)
2 Director-General's Proposed Strategy of International Agricultural Adjustment. C 75/18, August 1975. (Director-General's report to the 18th Session of the FAO Conference, November 1975.)
3 Agricultural Adjustment in Developed Countries. C 73/16, September 1973. (Revision of original report ERC: 72/3, prepared in cooperation with the Economic Commission for Europe under the direction of Mr Paul Lamartine Yates, 1972.)

The following case studies of various countries and selected issues relating to agricultural adjustment have been prepared under the direction of Dr James P. O'Hagan, Chief, Global Perspective Studies Service, Policy Analysis Division, FAO:

1 Agricultural Adjustment in the European Economic Community, with special reference to the Federal Republic of Germany. C 73/LIM/1, November 1973.
2 Sweden: A Case Study in Agricultural Adjustment. C 73/LIM/2, November 1973.
3 International Agricultural Adjustment: A Case Study of Japan. C 73/LIM/3, November 1973.
4 International Agricultural Adjustment: A Case Study of the United States. C 73/LIM/4, November 1973.
5 International Agricultural Adjustment: A Case Study of Sudan. C 73/LIM/5, August 1974.
6 Indonesia: A Case Study in Agricultural Adjustment. C 73/LIM/6, November 1973.
7 U.S.S.R.: A Case Study in Agricultural Adjustment. C 73/LIM/7, November 1973.

8 Central America: A Case Study in Agricultural Adjustment. C 73/LIM/8, October 1973.

9 Agricultural Protection: Domestic Policy and International Trade. C 73/LIM/9, November 1973.

10 Report on the Activities of the Diversification Fund of the International Coffee Organization. C 73/LIM/10, November 1973. (Contributed by the International Coffee Organization.)

11 Processed Agricultural Products and Agricultural Adjustment. C 73/LIM/11, November 1973. (Contributed by the OECD.)

12 International Agricultural Adjustment: Self-Sufficiency in Food and Food Commodities. C 75/LIM/1, October 1975.

13 Agricultural Protection and Stabilization Policies: A Framework of Measurement in the Context of Agricultural Adjustment. C 75/LIM/2, October 1975.

14 International Agricultural Adjustment: Case Study of Poland. C 75/LIM/3, August 1975.

15 International Agricultural Adjustment: Case Study of Sudan. C 75/LIM/4, November 1975.

# Index

GPSR Compliance
The European Union's (EU) General Product Safety Regulation (GPSR) is a set of rules that requires consumer products to be safe and our obligations to ensure this.

If you have any concerns about our products, you can contact us on

ProductSafety@springernature.com

In case Publisher is established outside the EU, the EU authorized representative is:

Springer Nature Customer Service Center GmbH
Europaplatz 3
69115 Heidelberg, Germany

www.ingramcontent.com/pod-product-compliance
Ingram Content Group UK Ltd.
Pitfield, Milton Keynes, MK11 3LW, UK
UKHW040226290926
15471UKWH00001B/24
* 9 7 8 1 3 4 9 0 3 4 7 8 9 *